Sedimentary Rocks in the Field

FOURTH EDITION

The Geological Field Guide Series

Barnes, J.W. and Lisle, R.J. (2004) *Basic Geological Mapping*, 4th edn. ISBN: 978-0-470-84986-6, 5th edn publishing (2011). ISBN: 978-0-470-68634-8

Fry, N. (1991) *The Field Description of Metamorphic Rocks*. ISBN: 978-0-471-93221-5

McClay, K.R. (1991) *The Mapping of Geological Structures*. ISBN: 978-0-471-93243-7

Milsom, J. and Eriksen, A. (2010) *Field Geophysics*, 4th edn. ISBN: 978-0-470-74984-5

Tucker, M.E. (2011) *Sedimentary Rocks in the Field*, 4th edn. ISBN: 978-0-470-68916-5

Sedimentary Rocks in the Field:

A Practical Guide

FOURTH EDITION

Maurice E. Tucker

Department of Earth Sciences

Durham University

Durham DH1 3LE

England

WILEY-BLACKWELL

A John Wiley & Sons, Ltd., Publication

Library of Congress Cataloguing-in-Publication Data
Tucker, Maurice E.
  Sedimentary rocks in the field : a practical guide / Maurice E. Tucker. – 4th ed.
      p. cm.
  Includes bibliographical references and index.
  ISBN 978-0-470-68916-5 (paper)
  1.   Sedimentary rocks. 2.   Petrology–Fieldwork.   I. Title.
  QE471.T84 2011
  552′.5–dc22

                                                                    2010033307

A catalogue record for this book is available from the British Library.

This book is published in the following electronic formats:
ePDF: 978-0-470-97368-4

Typeset in 9.5/11.5pt Times by Laserwords Private Limited, Chennai, India.
Printed and bound in Singapore by C.O.S. Printers Pte Ltd

7   2015

CONTENTS

PREFACE

The study of sedimentary rocks is often an exciting, challenging, rewarding and enjoyable occupation. However, to get the most out of these rocks, it is necessary to undertake precise and accurate fieldwork. The secret of successful fieldwork is a keen eye for detail and an enquiring mind; knowing what to expect and what to look for are important, although you do need to remain open-minded. Be observant, see everything in the outcrop, then think about the features seen and look again. This book is intended to show how sedimentary rocks are tackled in the field, and has been written for those with a geological background of at least first-year university or equivalent.

At the outset, this book describes how the features of sedimentary rocks can be recorded in the field, particularly through the construction of graphic logs. The latter technique is widely used since it provides a means of recording all details in a handy form; further, from the data, trends through a succession and differences between horizons readily become apparent. In succeeding chapters, the various sedimentary rock-types, textures and structures are discussed as they can be described and measured in the field. A short chapter deals with fossils since these are an important component of sedimentary rocks and much useful information can be derived from them for palaeoenvironmental analysis; they are also important in stratigraphic correlation and palaeontological studies. Having collected the field information, there is the problem of knowing what to do with it. A concluding section deals briefly with facies identification and points the way towards facies interpretations, and the identification of sequences and cycles.

Maurice E. Tucker

ACKNOWLEDGEMENTS

I should like to thank the friends and colleagues who have willingly read drafts of this handbook, suggested changes for this fourth edition, and kindly provided photographs, in particular Jenny Bevan, Telm Bover-Arnal, Jim Gallagher, Annette George, Dougal Jerram, Juan-Carlos Laya, Mike Mawson, Zahra Seyedmehdi and Paul Wright. Zoe took the picture on the front cover. I am indebted to Vivienne for providing support, encouragement, assistance and patience in ways that only a wife can.

1

INTRODUCTION

This book aims to provide a guide to the description of sedimentary rocks in the field. It explains how to recognise the common lithologies, textures and sedimentary structures and how to record and measure these features. There is a chapter explaining how fossils can be studied in the field, since they are common in many sedimentary rocks and are very useful for palaeoenvironmental analysis. A concluding chapter gives a brief introduction to the interpretation of sedimentary rock successions: facies, facies associations, cyclic sediments and sequences.

1.1 Tools of the Trade

Apart from a notebook (popular size around 20×10 cm), pens, pencils, appropriate clothing, footwear and a rucksack, the basic equipment of a field geologist comprises a hammer, chisel, hand-lens, compass-clinometer, tape measure, acid bottle, sample bags and marker pen. A Global Positioning System (GPS) receiver is most useful, and not only in remote areas (see Section 1.3). A hard hat for protection when working below cliffs and in quarries, and safety goggles for the protection of the eyes while hammering, should also be taken into the field and used; see Section 1.4 for further safety considerations. A camera is invaluable. Topographic and geological maps should also be carried, as well as any pertinent literature. If a lot of graphic logging is anticipated (see Section 2.4) then pre-prepared sheets can be taken into the field. Non-geological items that are useful and can be carried in a rucksack include a whistle, first-aid equipment, matches, emergency rations, knife, waterproof clothing and a 'space blanket'.

For most sedimentary rocks, a geological hammer of around 0.5–1 kg (1–2 lb) is sufficiently heavy. However, do be sympathetic to the out-crop and remember that many future generations of geologists will want to look at the exposure. In many cases it will not be necessary to hammer

Sedimentary Rocks in the Field: A Practical Guide, Fourth Edition   Maurice E. Tucker
© 2011 John Wiley & Sons, Ltd

since it will be possible to collect loose fresh pieces from the ground. A range of chisels can be useful, if a lot of collecting is anticipated.

A *hand-lens* is an essential piece of equipment; ×10 magnification is recommended since then grains and features down to 100 μm across can be observed. When holding the lens close to your eye, the field of view being examined with a ×10 lens is about 10 mm in diameter. To become familiar with the size of grains as seen through a hand-lens, examine the grains against a ruler graduated in millimetres. For limestones, it can be easier to see the grains after licking the freshly broken surface.

A compass-clinometer is important not only for taking routine dip and strike and other structural measurements, but also for measuring palaeocurrent directions. Do not forget to correct the compass for the angle between magnetic north and true north. This angle of declination is normally given on topographic maps of the region. You should also be aware that power lines, pylons, metal objects (such as your hammer) and some rocks (although generally mafic-ultramafic igneous bodies) can affect the compass reading and produce spurious results. A tape or steel rule, preferably several metres in length, is necessary for measuring the thickness of beds and dimensions of sedimentary structures. A metre-long staff with graduations can be useful for graphic logging. A compass usually has a millimetre-centimetre scale, which can be useful for measuring the size of small objects such as pebbles and fossils.

For the identification of calcareous sediments a plastic bottle of hydrochloric acid (around 10%) is useful, and if some Alizarin red S is added, then dolomites can be distinguished from limestones (limestones stain pink, dolomites no stain). Polythene or cloth bags, for samples, and a marker pen (preferably with waterproof, quick-drying ink), for writing numbers on the specimens, are also necessary. Friable specimens and fossils should be carefully wrapped to prevent breakage.

For *modern sediments* and *unconsolidated rocks* you will need a trowel and/or spade. A length (0.5–1.5 m) of clear plastic pipe, 5–10 cm in diameter, is very useful for pushing into modern sediments to obtain a crude core. Epoxy-resin cloth peels can be made in the field of vertical sections through soft sandy sediments. The techniques for taking such peels are given in Bouma (1969). In essence, cut and trim a flat vertical surface through the sediments; spray an epoxy resin onto the cut surface; place a sheet of muslin or cotton against the sediments and then spray the cloth. Leave the resin to set (~10 minutes) and then carefully remove the cloth. A thin layer of sediment should be glued to the cloth and will show the various structures. Lightly brush or shake off excess, unglued sediment. Modern beach, dune, river, tidal flat and desert sediments are ideal for treatment in this way. Fibre-glass foam (a hazardous substance, though) can also be sprayed onto unlithified sediment to take a sample.

1.2 Other Tools for the Field

More sophisticated instruments are taken into the field on occasion to measure a particular attribute of sedimentary rocks. Normally these are used as part of a more detailed and focused research effort, rather than during a routine sedimentological study. Such tools include the mini-permeameter, magnetic susceptibility recorder (kappameter), gamma-ray spectrometer, and equipment for ground-penetrating radar surveys and laser scanning (LIDAR, Light Detection and Ranging).

The mini-permeameter is a tool for estimating the permeability of a rock, and portable ones are available for use in the field.

The magnetic susceptibility of sedimentary rocks can be measured relatively easily in the field, although in many cases it is very weak. Mudrocks and others with high organic matter contents and iron minerals tend to give higher readings. With measurements taken every few centimetres or so, a mag-sus stratigraphy can be produced; from this, cycles and rhythms can be recognised, especially in basinal facies.

Gamma-ray spectrometry is a technique for measuring the natural gamma radiation emitted from rocks; it can be used to determine the amount of clay in a succession and so is useful for distinguishing different types of mudrock, or the variation in clay content of muddy sandstones and limestones. Measurement of the gamma-ray spectrum in the field with a portable spectrometer has been used to correlate surface outcrops with each other and with the subsurface.

Ground-penetrating radar is a useful technique for looking at the structure and variation of shallow subsurface sediments, as in modern floodplains and coastal plains. Sedimentary units, such as point-bar sands and ox-bow lake fills, can be recognised beneath the surface.

Laser scanning of an outcrop can produce a highly precise digital image that can be used in 3-D imaging software packages. Very accurate measurements can be taken from the data, for detailed study of such features as fracture orientation and bed thickness. The equipment is very expensive, however. Further information on an area can be obtained from multispectral remote sensing and from digital elevation models (DEMs, also called digital terrain models, DTMs), where the relief of an area can be revealed and so examined more closely. Using these techniques 3-D digital images of geological formations can be created.

1.3 Use of GPS (Global Positioning System) in Sedimentary Studies

GPS is becoming a standard instrument to take into the field for determining location, but it can also be used for measuring sedimentary sections. GPS provides a very precise location of where you are, and this reading of latitude and longitude (or grid reference) can be entered in your field notebook and on logging sheets. The receiver also enables you to travel from one place to another or to find a specified location, or to back-track from whence you came. The accuracy of the reading from the receiver depends on several factors (make and model, time at location, design, corrections, etc.) and the method of positioning. With autonomous GPS the precision is 5–30 m; using differential Global Positioning System (DGPS) and applying corrections, accuracy can be less than 3 m. However, a reference station does have to be set up for DGPS.

GPS receivers now have a good memory so that all readings of the day, indeed of the week, can be recalled so you can retrace your steps and visit the localities again with no difficulty. This is immensely useful in landscapes where there are no distinctive features. Data from the receiver can of course be downloaded directly to a PC so that a permanent record is kept of where you went.

Apart from the advantage of knowing where you are, GPS does have the potential to enable you to measure medium to large structures with a better degree of accuracy than by using just a map and tape measure. With channel-fills, reef bodies or conglomeratic lenses, etc., several 100 m across or more, taking several GPS measurements will enable a better estimate of dimension to be obtained. A laser range-finder can be useful for precise measurements of distant objects, such as features on cliff faces, and for knowing more precisely how far you are from somewhere.

1.4 Safety in the Field and General Guidance for Fieldwork

Working in the field should be a safe, enjoyable and very rewarding experience, as long as a few basic and sensible precautions are taken. Geological fieldwork is an activity involving some inherent risks and hazards, for example in coastal exposures, quarries, mines, river sections and mountains. Severe weather conditions may also be encountered in any season, especially in mountainous areas or at the coast. Fieldwork does involve an important element of self-reliance and the ability to cope alone or in a small group. You are responsible for your own safety in the field but, nevertheless, there are some simple precautions you can take to avoid problems and minimise risks.

- Do wear adequate clothing and footwear for the type of weather and terrain likely to be encountered. Try to know the weather forecast for the area before you go out for the day. Keep a constant lookout for changes. Do not hesitate to turn back if the weather deteriorates.
- Walking boots with good soles are normally essential. Sports shoes are unsuitable for mountains, quarries and rough country.
- Plan work carefully, bearing in mind your experience and training, the nature of the terrain and the weather. Be careful not to overestimate what can be achieved.
- Learn the mountain safety and caving codes, and in particular know the effects of exposure. All geologists should take a course in First Aid.
- It is good practice before going into the field to leave a note and preferably a map showing expected location of study and route, and time of return, with the people with whom you are staying/living.
- Know what to do in an emergency (e.g. accident, illness, bad weather, darkness). Know the international distress signal: six whistle blasts, torch flashes or waves of a light-coloured item of clothing, repeated at minute intervals.
- Carry at all times a small first-aid kit, some emergency food (chocolate, biscuits, mint cake, glucose tablets), a survival bag (or large plastic bag), a whistle, torch, map, compass and watch.
- Wear a safety helmet (preferably with a chin strap) when visiting old quarries, cliffs, scree slopes, caves and so on, or wherever there is a risk from falling objects. It is obligatory to do so when visiting working quarries, mines and building sites, along with a high-viz jacket and strong boots.

- Avoid hammering where possible, to be a conservationist.
- Wear safety goggles (or safety glasses with plastic lenses) for protection against flying splinters when hammering rocks or using chisels.
- Do not use one geological hammer as a chisel and hammer it with another; use only a soft steel chisel.
- Avoid hammering near another person or looking towards another person hammering. Do not leave rock debris on the roadway or verges.
- Be conservation-minded and have a sympathetic regard for the countryside and the great outdoors, and the people, animals and plants that live there.
- When you are collecting specimens, do not strip or spoil sites where special fossils and rare minerals occur. Only take what you really need for further work.
- Take special care near the edges of cliffs and quarries, or any other steep or sheer rock faces, particularly in gusting winds.
- Ensure that rocks above are safe before venturing below. Quarries with rock faces loosened by explosives are especially dangerous.
- Avoid working under an unstable overhang.
- Avoid loosening rocks on steep slopes.
- Do not work directly above or below another person.
- Never roll rocks down slopes or over cliffs for amusement.
- Do not run down steep slopes.
- Beware of landslides and mudflows occurring on clay cliffs and in clay-pits, or rockfalls from any cliffs.
- Avoid touching any machinery or equipment in quarries, mines or building sites. Comply with safety rules, blast warning procedures and any instructions given by officials. Keep a sharp lookout for moving vehicles, and so on. Beware of sludge lagoons.
- Do not climb cliffs, rock faces or crags, unless you are an experienced climber and have a partner.
- Take great care when walking or climbing over slippery rocks below high-water mark on rocky shores. More accidents to geologists, including fatalities, occur along rocky shorelines than anywhere else; keep an eye out for rogue waves.
- Beware of traffic when examining road cuttings.
- Railway cuttings and motorways are generally not open to geologists, unless special permission has been obtained from appropriate authorities.

- Do not enter old mine workings or cave systems unless you are experienced and properly equipped.
- Avoid getting trapped by the tide on intertidal banks or below sea-cliffs. Obtain local information about tides and currents. Pay particular attention to tidal range. For sea-cliffs, local advice can be obtained from the local coastguard's office.
- *Permissions*: Always try to obtain permission to enter private property. There are also many areas, such as national parks and nature reserves in many countries, and sites of special scientific interest (SSSIs) or protected sites, where official permission is required to collect samples, in some cases even just to carry out field observations and scientific study.
- *Risk assessment*: In many cases these days, it is necessary to conduct a risk assessment before commencing fieldwork. This may be a requirement for insurance purposes, as well as for research project proposals and permissions. Although perhaps seemingly unnecessarily bureaucratic, making a risk evaluation can be a useful exercise in thinking about the problems and issues you might encounter and so make you better prepared.

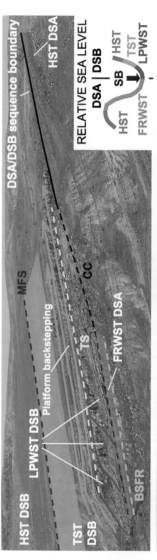

Figure 1.1 *Sequence stratigraphic interpretation of two middle Cretaceous sequences (DSA and DSB) from the Maestrat Basin, eastern Spain. Abbreviations: ST systems tract; T transgressive; H highstand; FRW forced regressive wedge; LPW lowstand prograding wedge; T transgressive; H highstand; FRW forced regressive wedge; BSFR basal surface of forced regression; CC correlative conformity; TS transgressive surface. See Bover-Arnal et al. (2009) for further information.*

2
FIELD TECHNIQUES

2.1 What to Look For

There are six aspects of sedimentary rocks to consider in the field, and these should be recorded in as much detail as possible. These are:

1. the **lithology**, that is the composition and/or mineralogy of the sediment;
2. the **texture**, referring to the features and arrangements of the grains in the sediment, of which the most important aspect to examine in the field is the grain-size and its variation;
3. the **sedimentary structures**, present on bedding surfaces and under-surfaces, and within beds, some of which record the palaeocurrents that deposited the rock;
4. the **colour** of the sediments;
5. the **geometry** and **relationships** of the beds or rock units to each other, and their lateral and vertical changes in thickness and composition; and
6. the nature, distribution and preservation of **fossils** contained within the sedimentary rocks.

A general scheme for the study of sedimentary rocks in the field is given in Table 2.1, starting from the larger-scale to the smaller-scale features.

The various attributes of a sedimentary rock combine to define a *facies*, which is the product of a particular depositional environment or depositional process in that environment. Facies identification and facies analysis are the next steps after the field data have been collected. These topics are briefly discussed in Chapter 8; there are numerous books on facies and facies models. Nowadays, there is much interest

Sedimentary Rocks in the Field: A Practical Guide, Fourth Edition   Maurice E. Tucker
© 2011 John Wiley & Sons, Ltd

Table 2.1 Broad scheme for the study of sedimentary rocks in the field, together with reference to appropriate chapters in this book.

Take a look at the locality from a distance and see the general arrangement of rock types and units. Make a quick assessment of any structural features, faults and folds, and any prominent sedimentological structures – channels, reefal limestones, clinoforms, synsedimentary folds, before having a closer look at the exposure

Record details of the locality and succession by means of notes and sketches in the field notebook, and photos; if appropriate, make a graphic log; if rocks are folded check way-up of strata; see Table 2.2

Identify lithology by establishing mineralogy/composition of the rock; see Chapter 3

Examine texture of the rock: grain-size, shape and roundness, sorting, fabric and colour; see Chapter 4

Look for sedimentary structures on bedding planes and bed undersurfaces, and within beds; see Chapter 5

Record the geometry of the sedimentary beds and units; determine the relationships between them and any packaging of beds/units or broad vertical grain-size/lithological/bed thickness changes; is the succession cyclic? See Sections 5.7 and 8.4

Search for fossils and note types present, modes of occurrence and preservation; see Chapter 6

Measure all structures giving palaeocurrent directions; see Chapter 7

Consider, perhaps at a later date, the lithofacies, cycles, sequences, depositional processes, environmental interpretations and palaeogeography; see Chapter 8

Undertake laboratory work to confirm and extend field observations on rock composition/mineralogy, texture, structures, fossils, etc.; pursue other lines of enquiry such as the biostratigraphy, diagenesis and geochemistry of the sediments and read the relevant literature, for example sedimentology/sedimentary petrology textbooks and appropriate journals, or use the internet. See Recommended Reading and chapter references in the broader-scale aspects of sedimentary successions: the geometric arrangements of rock units; the lateral and vertical variation in such features as lithology and grain-size; the packaging and stacking patterns of units; and the presence of depositional cycles in the succession. These aspects are treated in Sections 5.7 and 8.4. These features reflect the longer-term, larger-scale controls on deposition, primarily relative sea-level change, *accommodation* (the space available for sediment to be deposited), tectonics, sediment supply/sediment production and climate.

2.2 The Approach

The question of how many exposures to examine per square kilometre depends on the aims of the study, the time available, the lateral and vertical facies variation, and the structural complexity of the area. If it is a reconnaissance survey of a particular formation or group then well-spaced localities will be necessary. If a specific member or bed is being studied then all available outcrops will need to be looked at; individual beds may have to be followed laterally.

The best approach at outcrops is initially to survey the rocks from a distance, noting the general relationships and any folds or faults that are present. Some larger-scale structures, such as channels and erosion surfaces, the geometry of sedimentary rock units, bed thickness variations and the presence of cycles, are best observed from a short distance. Notice the way the rocks are weathering out and the vegetation. These may be reflecting the lithologies (e.g. mudrocks less well exposed or covered in vegetation) and may show the presence of cycles. Then take a closer look and see what lithologies and lithofacies are exposed there. In folded or vertical rocks, check the way-up of the strata using sedimentary structures such as cross-bedding, graded bedding, scours, sole structures, geopetals in limestone or cleavage-bedding relationships, so you know the younging direction (see Section 2.9).

Having established approximately what the outcrop has to offer, decide whether the section is worth describing in detail. If it is, it is best to record the succession in the form of a graphic log (see Section 2.4). If the exposure is not good enough for a log, then notes and sketches in the field notebook will have to suffice. In any event, not all the field information can go on the log.

11

2.3 Field Notes

Your notebook should be kept as neat and well-organised as possible. The location of the section being examined should be given precisely, preferably with a grid reference and possibly a sketch map too, so you can find it again in years to come. If you have GPS, this can give you a very precise location (see Section 1.3). You may wish to number your localities sequentially and put the numbers on a topographic map. You could use the pin-hole method (not favoured by everybody!); make a hole in the map with a pin and write the locality number on the back. Relevant stratigraphic information should also be entered in the notebook if you know it: formation name, age and so on. It is easy to forget such things with the passage of time. Incidental facts could be jotted down, such as the weather or a bird seen, to make the notebook more interesting and jolt the memory about the locality when looking back through the book in years to come.

Notes written in the field book should be factual, accurately describing what you can see. Describe and measure where possible the size, shape and orientation of the features as discussed and explained in later chapters of this book. Also record the structural data if the rocks are dipping or there are folds and cleavage present. Note major joints and fractures and their orientation, and any mineralisation. Make neat and accurate labelled sketches of features, with a scale, and orientation, such as direction of north.

Record the location and subject of *photographs* in the notebook. When taking photographs do not forget to put in a scale. Photomosaics of cliffs and quarries can be very useful for extensive exposures, and they can always be annotated directly or with an overlay.

One attribute of sediments that cannot be recorded adequately on a graphic log is the geometry of the bed or the rock unit as a whole (see Section 5.7). Sketches, photographs and descriptions should be made of the shape and lateral changes in thickness of beds as seen in quarry and cliff faces. Binoculars can be very useful for observing inaccessible cliffs and as a preliminary to closer examination. Local detailed mapping and logging of many small sections may be required in areas of poor exposure to deduce lateral changes. GPS can be useful here to get accurate locations of outcrops and even to get the dimensions of features. Where there are excellent outcrops and you have a focused research project, laser scanning a cliff or quarry face could help enormously.

Table 2.2   *The main points to be covered in a field notebook entry.*

Locality details: location, locality number, grid/GPS reference; date and time; weather

Stratigraphic horizon and age of rock unit, structural observations (dip, strike, cleavage, etc.)

The large-scale features of the exposure (e.g. faults, folds, channels, strong erosion surfaces, any geometric arrangements of units – thinning out, onlapping)

Lithology/mineralogy and texture: identify and describe/measure

Sedimentary structures: describe/measure and make sketches/take photographs

Palaeocurrent measurements: collect readings and plot rose diagram

Fossils: identify and make observations on assemblages, orientation, preservation, and so on

Construct graphic log if appropriate and sketches of lateral relationships

Note location of samples and fossils collected

Identify facies present, note facies associations and repetitions

Determine/measure rock units and any cycles in the succession

Make appropriate interpretations and notes for future work (e.g. in the lab)

Table 2.2 gives a checklist of the main points to be covered in the description of a locality in a field notebook.

2.4 Graphic Logs

The standard method for collecting field data of sediments and sedimentary rocks is to construct a graphic log of the succession (Figure 2.1). Logs immediately give a visual impression of the section, and are a convenient way of making correlations and comparisons between equivalent sections from different areas. Repetitions of facies, sedimentary cycles and general trends may become apparent, such as a systematic upward change in bed or cycle thickness or in grain-size, increasing or decreasing upward. In addition, the visual display of a graphic log helps with the interpretation of the succession. However, a log does emphasise the vertical changes in the succession, at the expense of lateral variations.

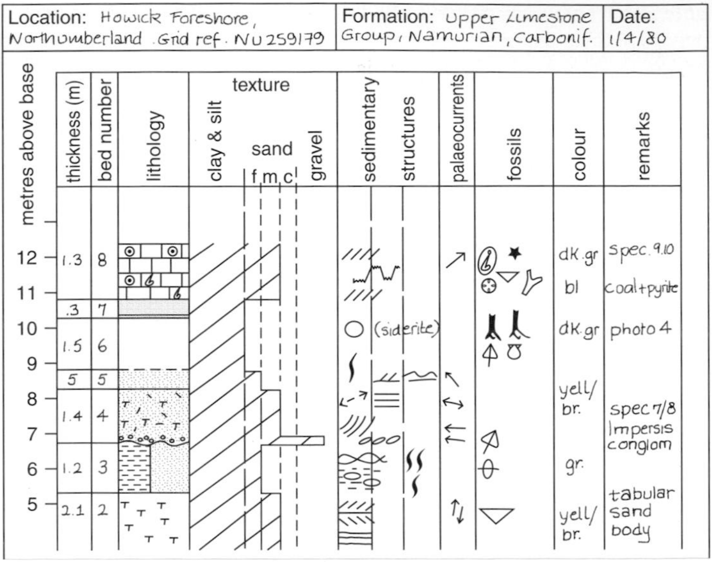

Figure 2.1 *An example of a graphic log; symbols are given in Figure 2.2.*

The vertical scale to use depends on the detail required, sediment variability and time available. For precise work on short sections, 1:10 or 1:5 is used, but for many purposes 1:50 (that is 1 cm on the log equals 0.5 m) or 1:100 (1 cm equals 1 m) is adequate. In some situations, it may not be necessary to log the whole succession, or the whole succession at the same scale. A representative log may be sufficient.

There is no set format for a graphic log; indeed, the features that can be recorded do vary from succession to succession. Features that it is necessary to record and that therefore require a column on the log are: bed or rock-unit thickness; lithology; texture, especially grain-size; sedimentary structures; palaeocurrents; colour and fossils. The nature of bed contacts can also be marked on the log. A further column for special or additional features ('remarks') can also be useful. Several types of graphic log form are illustrated by Graham in Tucker (1988), and the columns for a logging sheet are shown on the inside cover of this book and are available for downloading from www.wiley.com/go/sedimentaryrocks4e. If you are going to spend some time in the field then it is worth preparing the log sheets before you go. An alternative is to construct a log in your field notebook, but this is usually less satisfactory since the page size of most notebooks is too small.

Where exposure is continuous or nearly so, then there is no problem concerning the line of the log; simply take the easiest path. If the outcrop is good but not continuous everywhere it may be necessary to move laterally along the section to find outcrops of the succeeding beds. Some small excavations may be required where rocks in the succession, commonly mudrocks, are not exposed; otherwise enter 'no exposure' on the log. It is best to log from the base of the succession upwards. In this way you are recording how deposition changed as time progressed, rather than back through time, and it is generally easier to identify bed boundaries and facies changes by moving up the section.

2.4.1   Bed and rock-unit thicknesses

These are measured with a tape measure; care must be exercised where rocks dip at a high angle and the exposure surface is oblique to the bedding. Attention needs to be given to where boundaries are drawn between units in the succession; if there are obvious bedding planes or changes in lithology then there is no problem. Thin beds, all appearing identical, can be grouped together into a single lithological unit on the log, if a large scale is being used. Where there is a rapid alternation of thin beds of different lithology, for example interbedded sandstones and shales (heterolithics), they can be treated as one unit and notes made of the thicknesses and character of individual beds, noting any increases or decreases in bed thickness up through the unit.

Thus, when first approaching a section for logging, stand back a little and see where the natural breaks come in the succession to define the various beds or rock units.

It is often useful to give each bed or rock unit a number so as to facilitate later reference; begin at the stratigraphically lowest bed.

2.4.2   Lithology

On the graphic log, lithology is recorded in a column by using an appropriate ornamentation (Figure 2.2). If it is possible to subdivide the lithologies further, then more symbols can be added, or coloured pencils used. If two

15

SEDIMENTARY ROCKS IN THE FIELD

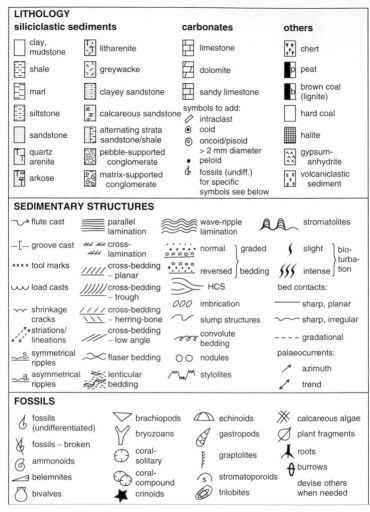

Figure 2.2 Symbols for lithology, sedimentary structures and fossils for use in a graphic log. This figure is on the Internet at: www.wiley.com/go/sedimentaryrocks4e.

lithologies are thinly interbedded, then the column can be divided in two by a vertical line and the two types of ornament entered. More detailed comments and observations on the lithology should be entered in the field notebook, reference to the bed or rock unit being made by its number.

2.4.3 Texture and grain-size

On the log there should be a horizontal scale for the textural column. For many rocks this will show mud (clay + silt), sand (divided into very fine, fine, medium, coarse, very coarse) and gravel. Gravel can be divided further if coarse sediments are being logged. To aid the recording of grain-size (or crystal-size), fine vertical lines can be drawn for each grain-size class boundary. Having determined the grain-size of a rock unit, mark this on the log and shade the area; the wider the column, the coarser the rock. Ornament for the lithology and/or sedimentary structures can be added to this textural column. In many logs, lithology and texture are combined into one column.

Other textural features, such as grain fabric, roundness and shape, should be recorded in the field notebook, although distinctive points can be noted in the remarks column. Particular attention should be given to these features if conglomerates and breccias are in the succession (see Section 4.6).

For the graphic logging of carbonate rocks, it is useful to combine the lithology/texture columns and use the Dunham classification (Figure 2.3). Thus you could have columns for lime mudstone (M), wackestone (W), packstone (P) and grainstone (G); a column for boundstones (B) can be added if reef-rocks or stromatolites are present. If there are very coarse limestones, separate columns can be added for rudstones (R) and floatstones (F) (see Section 3.5.2).

2.4.4 Sedimentary structures and bed contacts

Sedimentary structures and bed contacts within the strata can be recorded in a column by symbols. Sedimentary structures occur on the upper and lower surfaces of beds as well as within them. Separate columns can be used for surface and internal sedimentary structures if they are both common. Symbols for the common sedimentary structures are shown in Figure 2.2. Measurements, sketches and descriptions of the structures should be made in the field notebook.

Note whether *bed boundaries* are (i) sharp and planar, (ii) sharp and scoured or (iii) gradational; each can be represented in the lithology

17

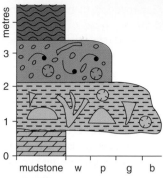

Figure 2.3 *Textural graphic log for limestones using the Dunham classification (w – wackestone, p – packstone, g – grainstone, b – boundstone).*

column by a straight, wavy/irregular or dashed line respectively. Types of bedding plane are shown in Figure 5.5.

It can be useful to have a column on a graphic log to show the degree of bioturbation. Estimate the bioturbation index (1 to 6) for a bed from Figure 5.69 and Table 5.8 (see Section 5.6.1).

2.4.5 Palaeocurrent directions

For the graphic log, readings can be entered either in a separate column or adjacent to the textural log as an arrow or trend line. The measurements themselves should be retained in the field notebook; make a table for the readings (see Figure 7.2).

2.4.6 Fossils

Fossils indicated on the graphic log should record the principal fossil groups present in the rocks. Symbols that are commonly used are shown in Figure 2.2. These can be placed in a fossil column alongside the sedimentary structures. If fossils make up much of the rock (as in some limestones) then the symbol(s) of the main group(s) can be used in the lithology column. Separate subcolumns on the textural log could be designated for rudstones and floatstones, where large fossils are abundant and in contact or in matrix-support fabric respectively (see Section 3.5.2). Observations on the fossils themselves should be entered in the field notebook (see Chapter 6).

2.4.7 Colour

The colour of a sedimentary rock is best recorded by use of a colour chart, but if this is not available then simply devise abbreviations for the colour column (see Section 4.8).

2.4.8 'Remarks' column

This can be used for special features of the bed or rock unit, such as degree of weathering (see Section 4.7) and presence of authigenic minerals (pyrite, glauconite, etc.) and supplementary data on the sedimentary structures, texture or lithology. The presence of joints and fractures can also be noted here (their spacing and density; see Section 5.5.8). You can also enter specimen numbers, the location of photographs taken, and cross-references to sketches in your notebook.

2.5 The Logging of Cores

The same graphic logging techniques and approaches applied in the field can be used for logging cores of subsurface sedimentary rocks, taken, for example, during exploration for hydrocarbon reservoirs, mineral deposits or just to establish the succession. The core logs also aim to give a representation of the grain-size/texture and of the lithology, presence of sedimentary structures, bioturbation index, fossils, and so on, through the succession, but in addition are a place for observations relating to the occurrence of porous zones and perhaps of oil or bitumen itself. The long-term grain-size/facies trends and the presence of cycles are keenly looked for, as is the occurrence of various bedding surfaces and discontinuities, which might indicate exposure/drowning, and so on. These features are not so easy to see in a 10 cm diameter core of course; and neither are macrofossils as abundant as in an outcrop. Measuring bed thickness in a core may not be so easy either, since nowadays many wells are drilled obliquely, or vertically then horizontally, and they may deviate all over the place. Fractures are often of great interest (in terms of poroperm) and may warrant special attention (see Section 5.5.8), but fractures may be induced during drilling.

Of particular importance in a core is the occurrence of faults, which of course will cut-out or repeat the succession. It can be difficult to recognise a fault in the first place and then very difficult to determine the throw, type and direction of movement, and so the significance of the fault, in a single core.

Cores are best examined if they have been cut in half so a flat surface can be seen. Cores are often dirty and not polished in any way, so a supply of water and a sponge, or a polythene spray bottle with water, are almost essential for wetting the surface to bring out the structures. Plenty of space for laying out the cores is also useful. Also needed are a hand-lens (or binocular microscope), hammer, steel point/penknife (to test for hardness), dilute HCl, grain-size chart (see Figure 4.1), sample bags and waterproof marker pen. It is often very useful if the wireline logs are also available for the well and can be looked at alongside the core itself or the core log.

The charts for cores are not really any different from those used in the field; the aim is the same, the recording of all necessary data for the project in hand. A column for oil staining can be useful – none, weak, strong. There may be zones of no core recovery (mark by crossed lines) and in places the rock may have disintegrated through the drilling so only comminuted rock is obtained. Core is precious material (since it is usually extremely expensive to obtain) so it is best always to save half of the core for future reference, rather than take a whole piece for chemical analysis or microfossil extraction. For further information on core logging see Blackbourne (2000). A logging sheet for core material available on the Internet at www.wiley.com/go/sedimentaryrocks4e.

2.6 Lithofacies Codes

For the study of certain sedimentary rock types – chiefly glacial, fluvial and deepwater clastic sediments – shorthand codes have been devised to make the description of outcrops and logging of sections quicker and more efficient. As an example, in one scheme (Table 2.3), G, S, F and D refer to gravels (conglomerates), sands (sandstones), fines (muds/mudrocks) and diamictons/diamictites (muddy gravels/conglomerates), respectively, and m, t, p, r, h, etc. are added as qualifiers, if the sediments are massive, trough cross-bedded, planar cross-bedded, rippled, horizontally laminated, etc. The letters vf, f, m, c or vc added before the S or G would refer to very fine, fine, medium, coarse or very coarse. Thus fShr would refer to a fine, horizontally laminated and rippled sandstone.

With carbonates (Table 2.4), the initials of grainstone (G), packstone (P), wackestone (W), mudstone (M), boundstone (B), and so on, can

Table 2.3 *Lithofacies codes for siliciclastic sediments. This shorthand notation can be used usefully with fluvial, glacial and deepwater sediments for rapid description, but be aware of the limitations of such a scheme (see Section 2.6).*

Lithologies
G – gravel, S – sand, F – fines (mud), D – diamicton
Qualifiers
m – massive, p – planar cross-bedded, t – trough cross-bedded,
   r – ripple cross-laminated, h – horizontal-laminated,
   l – laminated, r – rootlets, p – pedogenic, and so on
Prefixes
f – fine, m – medium, c – coarse

Table 2.4 *Lithofacies codes for carbonate sediments, useful for rapid description, but be aware of the limitiations of such a scheme (see Section 2.6).*

Lithologies
M – mudstone, W – wackestone, P – packstone, G – grainstone,
   B – boundstone, R – rudstone, F – floatstone, D – dolomite
Qualifiers
f – fenestral, s – stromatolitic, o – ooidal, p – peloidal,
   b – bioclastic, cr – crinoidal, v – vuggy, and so on
Prefixes
f – fine, m – medium, c – coarse, cx – crystalline,
   d – dolomitic, s – siliceous, and so on

be used along with appropriate qualifiers such as s (stromatolitic), f (fenestral), o (ooidal), c (coral), q (quartzitic), and so on. Thus, a fGqo would be a fine-grained quartzitic oolitic grainstone (see Section 3.5.2 for limestone types).

The lithofacies codes approach can be useful if you have a very thick succession of sediments to document. Devise your own abbreviations as appropriate, depending on the lithofacies and sedimentary structures present, with explanation in the field notebook.

Lithofacies codes have come in for some criticism, however; since they may lead to an over-simplification or generalisation of a succession. There is a danger in pigeon-holing a sediment with a code; the scheme

1. Introduction

2. Field Techniques

3. Sedimentary Rock Types

4. Sedimentary Rock Texture

has to be flexible to allow for the unusual, but perhaps environmentally significant, rock-type to be included appropriately.

2.7 Collecting Specimens

For much sedimentological laboratory work, samples of hand specimen-size are sufficient, although this does depend on the nature of the rock and the purpose for which it is required. Samples should be of in situ rock, and you should check that they are fresh, unweathered and representative of the lithology. If it is necessary to hammer the outcrop, wear safety goggles to protect your eyes.

Label the rock sample; give it (and its bag) a number using a water-proof felt-tip pen. In many cases, it is useful or necessary to mark the way-up of the specimen; an arrow pointing to the stratigraphic top is sufficient for this. For detailed fabric studies, the orientation of the rock (strike and dip direction) should also be marked on the sample. As a safeguard, specimen number and orientation data can be recorded in the field notebook, with a sketch of the specimen.

Specimens can be collected for the extraction of *microfossils*, such as foraminifera from Mesozoic-Cainozoic mudrocks and conodonts from Palaeozoic limestones. A hand-sized sample ($\sim 1\,kg$) is usually sufficient for a pilot study. *Macrofossils* too can be collected in the field, for later cleaning and identification. Faunas from different beds or lithofacies should be kept in separate bags. Many fossils will need to be individually wrapped in newspaper.

It is good practice to collect sparingly and not just for the sake of it; only take away what is really necessary for your project. Many fossils can be identified sufficiently in the field if the study is of the sedimentology and palaeoenvironments, and need not be taken home.

2.8 Presentation of Results

Once the field data have been collected it is often necessary to present or communicate this information to others. Summary graphic logs are very useful too, as are field sketches and photographs, and lithofacies maps. After the fieldwork, data collected can be entered into digital logs for permanent record. The usual drawing packages can be used (e.g. CorelDRAW®, FreeHand) or specific graphic log programs (e.g. SedLog in Zervas *et al.* (2009), Computers & Geosciences, 35, 2151–2159), which can be downloaded freely from the Internet at www.sedlog.com, a site hosted by Royal Holloway University of London.

A summary log may consist of one column depicting the grain-size, principal sedimentary structures and broad lithology; see the examples in Chapter 8 (e.g. Figures 8.15, 8.16, 8.18, 8.20 and 8.24). Such a log gives an immediate impression of the nature of the rock succession, especially the upward change in grain-size and lithology. If it is necessary to give more information, then lithology can be represented in a separate column alongside the log depicting the grain-size and structures (e.g. Figure 2.4).

The larger-scale patterns of grain-size change within a succession (i.e. upward-fining or upward-coarsening) are often of interest, as are the longer-term patterns of facies change, that is whether there is a long-term shallowing-up or deepening-up. These patterns can be indicated by long arrows or narrow triangles alongside the logs to show the trends (see Figure 8.9 for an example).

Line drawings of the lateral relationships of sedimentary rock units should be included in reports, along with sketches and/or photographs of more detailed aspects of the sedimentary story.

Maps showing the distribution of lithofacies of laterally equivalent strata over an area can be very useful. Maps can also be drawn to show

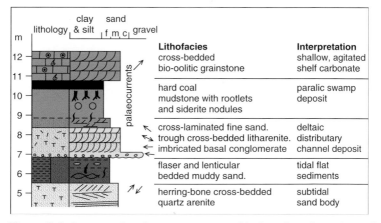

Figure 2.4 *An example of a summary graphic log, based on data of Figure 2.1.*

variations in specific features of the facies, such as sediment grain-size, thickness and sandstone/shale ratio. There are many computer programs available for processing field data and constructing logs, graphs and maps, which can impress the reader of a report. Statistical analysis can be undertaken of bed thickness data, palaeocurrent data and other measurements.

2.9 The Way-Up of Sedimentary Strata

Sedimentary rocks are commonly folded, and in small outcrops, especially where there are vertical beds, it may not be immediately apparent which is the top and which is the bottom of the succession. In these cases, it may be necessary to check which is the younging direction of the beds. The way-up of the strata can be deduced from many of the sedimentary structures described later in this book and shown in Figure 2.5.

Good structures to use are:

- **cross-bedding** – look for the truncations of the cross-beds;
- **graded bedding** – coarser grains at the base of the bed (although be aware of the possibility of inverse grading, especially in conglomerates and very coarse sandstones) (see Figure 5.37);
- **scours and channels** – sharp erosive bases to beds cutting down into underlying sediments, usually with coarser grains above the surface and finer grains below (see Figures 5.3 and 5.57a);
- **sole structures** – flute, groove and tool marks on the undersides of beds (see Figures 5.1 and 5.2);
- **mudcracks** – v-ing downward cracks with sand fills (see Figure 5.39);
- **dewatering and load structures** – flames, sedimentary dykes, sand volcanoes;
- **ripples and mudcracks** – generally occurring on the upper surfaces of beds;
- **cross-lamination** – look for truncations of the cross-laminae;
- **geopetal structures in limestones** – internal sediment in the lower part and calcite cement in the upper part of the cavity (see Figures 5.41 and 5.42);
- **certain trace fossils and fossils in growth position** (e.g. corals, rudistid bivalves, *Conichnus*).

In addition, there are some structural features that can be used to deduce way-up: *bedding-cleavage* relationships and *fold-facing* directions.

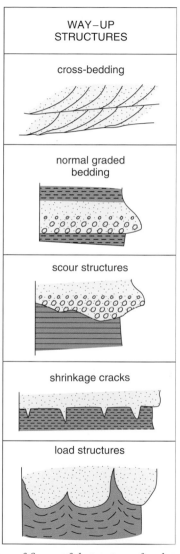

Figure 2.5 *Sketches of five useful structures for determining way-up of strata.*

2.10 Stratigraphic Practice

Stratigraphically, rocks are divided up on the basis of lithology (lithostratigraphy), fossils (biostratigraphy), key surfaces (sequence stratigraphy) and time (chronostratigraphy). From field studies, sedimentary rocks are primarily considered in purely descriptive lithostratigraphic terms, shown in Table 2.5.

2.10.1 Lithostratigraphy

The fundamental unit in lithostratigraphy is the *formation*, possessing an internal lithological homogeneity and serving as a basic mappable unit. Adjacent formations should be readily distinguishable on physical or palaeontological grounds. Boundaries may be gradational, but they should be clearly, even if arbitrarily, defined in a designated type section or sections. Although thickness is not a criterion, formations are typically a few to several 100 metres thick. Thickness will vary laterally over an area, and formations are commonly diachronous on a large scale. Stratigraphically adjacent and related formations, such as those deposited within the same basin, may be associated so as to constitute a *group* (typically 10^3 m thick). Two or more associated groups could be taken together to form a *supergroup*. A formation may be subdivided into *members*, characterised by more particular lithological features, and if there is a distinctive bed within a member this can be given a specific name. Lithostratigraphical units are given geographical names; with formations, a word to indicate the dominant lithology can be included.

There are several publications that give details of the procedure for defining lithostratigraphic units, and there is an International Code (see

Table 2.5 *Hierarchy of lithostratigraphic units.*

Supergroup – a formal assemblage of related or superposed groups
Group – an assemblage of formations
Formation – the fundamental lithostratigraphic unit, identified by lithological characteristics and stratigraphic position, generally tabular. Mappable at the Earth's surface and traceable into the subsurface. Several tens to hundreds of metres thick
Member – a formal lithostratigraphic unit constituting a formation
Bed – distinctive subdivision of a member; the smallest formal lithostratigraphic unit of sedimentary rock

the references in Recommended Reading). In many parts of the world, older stratigraphic names are in use that do not conform with the International Code. To erect your own lithostratigraphy for an area, draw up the stratigraphic succession for the area or region, and separate the various rock units on the basis of lithology and major unconformities/disconformities. Then choose a locality or area where the best outcrop of the particular lithology occurs, where all or the majority of the succession is exposed, and name the formation after this place. The formation may have several distinctive units within it that could be called *members*; also name these after their best outcrop. If you are working in a large area with rocks of quite different ages or tectonic setting, then you may have more than one group of sedimentary rocks present.

2.10.2   Sequence stratigraphy

Sequence stratigraphy is a methodology that provides a framework for understanding the evolution of depositional systems in space and time, as well as facilitating palaeogeographic reconstructions and providing a degree of prediction of facies away from known areas. Sequence stratigraphy is an increasingly popular way – but a sometimes contentious one, it has to be said – of dividing up the stratigraphic record on the basis of *key surfaces*, that is *unconformities* and their *correlative conformities*, and *flooding surfaces*, into *sequences*. A sequence is defined as 'a succession of strata deposited during a full cycle of change in accommodation or sediment supply' (Catuneanu *et al.*, 2009; also see other references in Recommended Reading for more information). Accommodation is the space available for sediments, positive (increasing) and negative (decreasing). An *unconformity* – taken as a *sequence boundary* (sb) in several of the sequence models – is a surface separating younger from older strata along which there is evidence of subaerial exposure; it will pass laterally (basinwards) into a *correlative conformity*. Major drowning surfaces have also been interpreted as sequence boundaries.

A sequence can usually be divided into several *systems tracts* (STs) (defined as a linkage of contemporaneous depositional systems, i.e. related facies or facies association), deposited during a specific part of a cycle of change in accommodation (relative sea level), that is falling stage systems tract (FSST, also called forced regressive systems tract, FRST); lowstand systems tract (LST); transgressive systems tract (TST) and highstand systems tract (HST) (see Figure 2.6). In one model the

HST + FSST + LST constitutes the regressive systems tract (RST). Apart from the unconformity-correlative conformity, other *key surfaces* are the *transgressive surface* (ts), which may be coincident with the subaerial unconformity in more proximal (landward) parts of a basin, at the base of the TST, and the *maximum flooding surface* (mfs), which separates the TST from the HST (Figure 2.6). In more distal parts of the basin, there is commonly a *condensed section* (CS) equivalent to the upper part of the TST, the mfs and the lower part of the HST. The common sequence stratigraphic terms are defined in Table 2.6. See Figure 1.1 for an example of a sequence stratigraphic interpretation of a Cretaceous carbonate outcrop.

There are currently several sequence stratigraphic models in use: the depositional sequence, genetic sequence and transgressive-regressive (T-R) sequence, each with its particular bounding surfaces and ST arrangements (Catuneanu *et al.*, 2009). You should use the model that is most appropriate for your succession. This depends on the data available, the depositional setting and facies: marine vs non-marine, clastic vs carbonate, shallow vs deep, and your viewpoint. As an example Figure 2.6 shows the basic model for a 4-ST depositional sequence.

Table 2.6 *Hierarchy of sequence-stratigraphic units.*

Sequence: succession of strata deposited during a full cycle of change in accommodation or sediment supply. Three types recognised: depositional sequence, genetic sequence, T-R sequence

Key surfaces: subaerial unconformity-correlative conformity, transgressive surface, maximum flooding surface, which divide sequences into **systems tracts**

Subaerial unconformity-correlative conformity: the subaerial unconformity forms under subaerial conditions as a result of fluvial erosion or bypass, pedogenesis, wind degradation, dissolution and/or karstification. The submarine equivalent, the correlative conformity, is a surface marking the change from highstand normal regression to forced regression/lowstand facies

Transgressive surface (ts) (also maximum regressive surface): stratigraphic surface that marks a change in stratal stacking patterns from lowstand normal regression to transgression

(continued overleaf)

Table 2.6 (*continued*)

Maximum flooding surface (mfs): marks a change in stratal stacking patterns from transgression to highstand normal regression, usually the deepest-water facies; distal areas starved of sediment form a **condensed section (CS)**, overlain by shallowing-upward succession

Sequence boundary (SB): separates one sequence from another; which key surface is the SB depends on the sequence model. In the depositional sequence model (Figure 2.6), the sequence boundary is the subaerial unconformity-correlative conformity

Systems tract (ST): a linkage of contemporaneous depositional systems
Five are commonly distinguished: (i) falling stage (FSST) (also called forced regressive, FRST), facies deposited during negative accommodation (falling relative sea level) and forced regression; (ii) lowstand (LST) – facies deposited during relative sea-level low and normal regression; (iii) transgressive (TST) – facies deposited during positive accommodation (rising relative sea level) and transgression; (iv) highstand (HST) – facies deposited during relative sea-level high and normal regression; and (v) regressive systems tract (RST) – HST + FSST + LST

Parasequence (psq): relatively conformable succession of genetically related beds or bedsets, typically metre-scale, which may be bounded by marine flooding surfaces

Parasequence set: succession of genetically related parasequences that have a distinctive stacking pattern (e.g. thinning up); usually bounded by major marine flooding surfaces

Marine flooding surface (fs): a surface that separates younger from older strata, across which there is evidence of an abrupt increase in water depth

In practice, there are features that you can unequivocally describe in the field (such as sediment geometries and relationships, grain-size and facies patterns and their vertical and lateral changes, bed and metre-scale cycle trends; see Section 8.4.5) and with these data you can then interpret the succession in terms of sequence stratigraphy and use the most appropriate model.

1. Introduction

2. Field Techniques

3. Sedimentary Rock Types

4. Sedimentary Rock Texture

SEDIMENTARY ROCKS IN THE FIELD

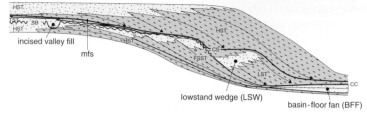

Figure 2.6 *A general sequence stratigraphic model (simplified) which can be applied to siliciclastic and carbonate sediments, showing the arrangement of systems tracts and key surfaces, and location of sands and muds. FSST, LST, TST, HST = falling stage, lowstand, transgressive and highstand systems tracts; IVF = incised valley fill; SB = sequence boundary and its correlative conformity, CC; ts = transgressive surface; mfs = maximum flooding surface; CS = condensed section.*

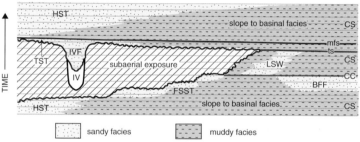

Figure 2.7 *Chronostratigraphic diagram for the succession shown in Figure 2.6. This diagram shows the distribution of sediment in time and space, and brings out the times of subaerial exposure. See caption to Figure 2.6 for abbreviations.*

Many sequences, especially in platform carbonates and shoreline-shelf clastics, are composed of several or many metre-scale cycles, termed *parasequences* (originally defined by *flooding surfaces* at their bases), and then the STs can be defined by the *stacking patterns* of the parasequences (e.g. whether they thin/thicken-up or fine/coarsen-up) and upward facies changes, which are a reflection of the cycle(s) of accommodation change. See Section 8.4 for further information on how to recognise the key surfaces and STs in the field, and what features to look for in successions of parasequences.

Sequences within an area are generally named by letters or numbers, or a combination of both, working from the base upwards.

2.10.3 Chronostratigraphy

Chronostratigraphy considers the stratigraphic record in terms of time. It can be very useful to think of strata in this way, especially when examining the succession on a basin-scale and there are breaks in sedimentation and periods of uplift. A chronostratigraphic diagram depicts the succession in space and time, and so does not indicate thickness. It will show where and when deposition and subaerial exposure took place, and bring out the relationships between different units.

Table 2.7 *The Cainozoic chronostratigraphical scale with approximate ages of the beginning of the series.*

System	Series	Stage	Ma
Quaternary	Holocene		0.1
	Pleistocene		1.7
Neogene	Pliocene	Gelasian	5.5
		Piacenzian	
		Zanclean	
	Miocene	Messinian	24
		Tortonian	
		Serravallian	
		Langhian	
		Burdigalian	
		Aquitanian	
Palaeogene	Oligocene	Chattian	34
		Rupelian	
	Eocene	Priabonian	54
		Bartonian	
		Lutetian	
		Ypesian	
	Palaeocene	Thanetian	65
		Selandian	
		Danian	

Ma, millions of years ago.

Table 2.8 *The Mesozoic chronostratigraphical scale with approximate ages of the beginning of the series.*

System	Series	Stage	Ma
Cretaceous	Upper	Maastrichtian	99
		Campanian	
		Santonian	
		Coniacian	
		Turonian	
		Cenomanian	
	Lower	Albian	142
		Aptian	
		Barremian	
		Hauterivian	
		Valanginian	
		Berriasian	
Jurassic	Upper (Malm)	Tithonian	156
		Kimmeridgian	
		Oxfordian	
	Middle (Dogger)	Callovian	178
		Bathonian	
		Bajocian	
		Aalenian	
	Lower (Lias)	Toarcian	200
		Pliensbachian	
		Sinemurian	
		Hettangian	
Triassic	Upper	Rhaetian	230
		Norian	
		Carnian	
	Middle	Ladinian	251
		Anisian	
	Lower	Olenekian	
		Induan	

Table 2.9 *The Palaeozoic chronostratigraphical scale with approximate ages of the beginning of the series.*

System	Series	Stage	Ma
Permian	Lopingian	Changhsingian	
		Wuchiapingian	
	Guadalupian	Capitanian	270
		Wordian	
		Roadian	
	Cisuralian	Kungurian	290
		Artinskian	
		Sakmarian	
		Asselian	
Carboniferous	Upper	Gzhelian	323
		Kasimovian	
		Moscovian	
		Bashkirian	
	Lower	Serpukhovian	360
		Viséan	
		Tournaisian	
Devonian	Upper	Fammenian	382
		Frasnian	
	Middle	Givetian	395
		Eifelian	
	Lower	Emsian	417
		Pragian	
		Lochkovian	
Silurian	Upper	Pridolian	424
		Ludlovian	
	Lower	Wenlockian	443
		Llandoverian	
Ordovician	Upper	Ashgillian	458
		Caradocian	
	Middle	Llanvirn	475
	Lower	Arenig	490
		Tremadocian	
Cambrian	Upper		500
	Middle		511
	Lower		545

Once a lithostratigraphic or sequence stratigraphic analysis has been completed, it is useful to sketch out the chronostratigraphy (Figure 2.7).

Chronostratigraphic divisions are time-rock units, that is they refer to the succession of rocks deposited during a particular interval of time. The chronostratigraphy of the Palaeozoic, Mesozoic and Cainozoic is shown in Tables 2.7–2.9 with the system, series and stage names and the approximate age of the beginning of the stage.

3
SEDIMENTARY ROCK TYPES

3.1 Principal Lithological Groups

For the identification of sedimentary rock-types in the field, the two principal features to note are composition-mineralogy and grain-size. On the basis of origin, sedimentary rocks can be grouped broadly into four categories (Table 3.1).

The most common lithologies are the sandstones, mudrocks and limestones (which may be altered to dolomites). Other types, evaporites, ironstones, cherts and phosphates, are rare or only locally well developed, and volcaniclastics are important in regions affected by volcanism.

In some cases you may have to think twice as to whether the rock is even sedimentary in origin or not. Greywacke sandstones, for example, can look very much like dolerite or basalt, especially in a hand-specimen away from the outcrop. Points generally indicating a sedimentary origin include the presence of:

1. stratification;
2. sedimentary structures on bedding surfaces and within beds;
3. fossils;
4. grains or pebbles that have been transported (i.e. clasts); and
5. specific minerals of sedimentary origin (e.g. glauconite, chamosite).

3.1.1 Terrigenous clastic rocks

These are dominated by detrital grains (silicate minerals and rock fragments especially) and include the sandstones, mudrocks, conglomerates and breccias.

Sandstones are composed of grains chiefly between 1/16 (0.06) and 2 mm in diameter (see Section 3.2). Bedding is usually obvious and sedimentary structures are common within the beds and upon the bedding surfaces.

Sedimentary Rocks in the Field: A Practical Guide, Fourth Edition   Maurice E. Tucker
© 2011 John Wiley & Sons, Ltd

Table 3.1 *The four principal categories of sedimentary rock with the broad lithological groups.*

Terrigenous clastics	Biochemical-biogenic-organic deposits	Chemical precipitates	Volcani-clastics
Sandstones, mudrocks, conglomer-ates + breccias	Limestones + dolomites, coal, phophorites, chert	Ironstones, evaporites	Composed of tephra (pyroclastic material), tuffs

Conglomerates and breccias (see Section 3.3), also referred to as rudites, consist of large clasts (pebbles, cobbles and boulders), more rounded in conglomerates, more angular in breccias, with or without a sandy or muddy matrix.

Mudrocks (see Section 3.4) are fine-grained particles mostly less than 1/16 (0.06) mm in diameter, and are dominated by clay minerals and silt-grade quartz. Many mudrocks are poorly bedded and also poorly exposed. Colour is highly variable, as is fossil content.

With increasing grain-size muds/mudrocks grade into sands/ sandstones and the latter into gravels/conglomerates, and there are also mixtures of all three of course. Figure 3.1 shows terms for mixtures of clay/silt/sand and mud/sand/gravel. Sediments consisting of a rapid interbedding of sandstones and mudrocks are often referred to as *heterolithic* facies.

3.1.2 Limestones and dolomites

Limestones (see Section 3.5) are composed of more than 50% $CaCO_3$ and so the standard test is to apply dilute acid (HCl) – limestone rock will fizz. Many limestones are a shade of grey, but white, black, red, buff, cream and yellow are also common colours. Fossils are commonly present, in some cases in large numbers.

Dolomites (also called dolostones) are composed of more than 50% $CaMg(CO_3)_2$. They react little with dilute acid (although a better fizz will be obtained if the dolomite is powdered first), but more readily with warm or more concentrated acid. Alizarin red S in hydrochloric acid stains limestone pink to mauve, whereas dolomite

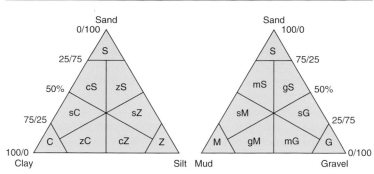

Figure 3.1 *Classification of sand (S), silt (Z) and clay (C) mixtures, where s = sandy, z = silty and c = clayey, and of sand (S), mud (M) and gravel (G) mixtures, where s = sandy, m = muddy and g = gravelly. For rocks, S is sandstone, M is mudrock, Z is siltstone and G is conglomerate and g is conglomeratic.*

is unstained. Many dolomites are creamy yellow or brown in colour and they are commonly harder than limestones. Most dolomites have formed by replacement of limestone, and in many cases as a result of this the original structures are poorly preserved. Poor preservation of fossils and the presence of *vugs* (irregular holes) are typical of dolomites.

3.1.3 Other lithologies

Gypsum is the only evaporite mineral (see Section 3.6) occurring commonly at the Earth's surface, mostly as nodules of very small crystals in mudrock, although veins of fibrous gypsum (satin spar) are usually associated. Evaporites such as anhydrite and halite are only encountered at the surface in very arid areas.

Ironstones (see Section 3.7) include bedded, nodular, oolitic and replacement types. They commonly weather to a rusty yellow or brown colour at outcrop. Some ironstones feel heavy relative to other sediments.

Cherts (see Section 3.8) are mostly cryptocrystalline to microcrystalline siliceous rocks, occurring as very hard bedded units or nodules in other lithologies (particularly limestones). Many cherts are dark grey to black, or red.

Sedimentary *phosphate* deposits, or *phosphorites* (see Section 3.9), mostly consist of concentrations of bone fragments and/or phosphate nodules. The phosphate itself is usually cryptocrystalline, dull on a fresh fracture surface with a brownish or black colour.

Organic sediments (see Section 3.10) such as hard coal, brown coal (lignite) and peat should be familiar and oil shale can be recognised by its smell and dark colour.

Volcaniclastic sediments (see Section 3.11), which include the tuffs, are composed of material of volcanic origin, chiefly lava fragments, volcanic glass and crystals. Volcaniclastics are variable in colour, although many are a shade of green through chlorite replacement. They are commonly badly weathered at outcrop. The term *pyroclastic* refers to material derived directly from volcanic activity whereas the term *epiclastic* is used to refer to 'secondary' sediments such as debris flow and fluvial deposits resulting from the reworking of pyroclastic material.

3.2 Sandstones

Sandstones are composed of five principal ingredients: rock fragments (lithic grains), quartz grains, feldspar grains, matrix and cement. The matrix consists of clay minerals and silt-grade quartz, and in most cases this fine-grained material is deposited along with the sand grains. It can form from the diagenetic breakdown of labile (unstable) grains, however, and clay minerals can be precipitated in pores during diagenesis. Cement is precipitated around and between grains, also during diagenesis; common cementing agents are quartz and calcite. Diagenetic hematite stains a sandstone red.

The composition of sandstone is largely a reflection of the geology and climate of the source area. Some grains and minerals are mechanically and chemically more stable than others. Minerals, in decreasing order of stability, are quartz, muscovite, microcline, orthoclase, plagioclase, hornblende, biotite, pyroxene and olivine. A useful concept is that of *compositional maturity*: immature sandstones contain many unstable grains (rock fragments, feldspars and mafic minerals). Mature sandstones consist of quartz, some feldspar and some rock fragments, whereas supermature sandstones consist almost entirely of quartz. In general, compositionally immature sandstones are deposited close to the source area, whereas supermature sandstones result from long-distance transport and much reworking. The minerals present in a sandstone thus

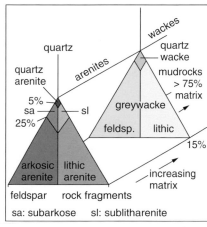

Figure 3.2 *Classification of sandstones. Careful use of a hand-lens in the field should enable recognition of the main sandstone types: quartz arenite, arkose, litharenite and greywacke.*

depend on the geology of the source area, the degree of weathering there and on the length of the transport path.

The accepted classification of sandstones is based on the percentages of quartz (+chert), feldspar, rock fragments and matrix in the rock (Figure 3.2). Sandstones containing an additional, non-detrital component, such as carbonate grains (ooids, bioclasts, etc.), are referred to as *hybrid sandstones* and are described in succeeding sections. The composition of a sandstone is based on a modal analysis determined from a thin-section of the rock using a petrological microscope and a point counter.

In the field, it is often possible to assess the composition and give the sandstone a name, through close scrutiny with a hand-lens. This can be verified later in the laboratory when a thin-section is available. With a hand-lens, attempt to estimate the amount of matrix present in a sandstone and thus determine whether it is an *arenite* (a clean sandstone) or a *wacke* (>15% matrix, a muddy sandstone).

The nature of the grains themselves is best determined from their fracture surface. The percentage charts in Figure 3.3 can be used to estimate the proportions of the various constituents present.

1. Introduction

2. Field Techniques

3. Sedimentary Rock Types

4. Sedimentary Rock Texture

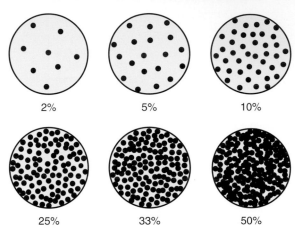

Figure 3.3 *Percentage estimation charts. Use for giving a rough estimate of the percentage of grains or bioclasts (fossils) or crystals (etc.) present in the rock.*

Quartz grains will appear milky to clear and glassy (Figure 3.4), without cleavage surfaces but with fresh conchoidal fractures. Quartz grains commonly have a quartz cement overgrowth around them and this will have flat crystal surfaces, which may catch the light (Figure 3.4). *Feldspar* grains are commonly slightly to totally replaced by clay minerals (Figure 3.4), so they do not have the fresh glassy appearance of quartz; they are usually white in colour, possibly pink. Cleavage surfaces and/or twin planes are usually visible on the fresh fracture surfaces, as they reflect the light. In many sandstones at outcrop, the feldspar grains have been dissolved out leaving a porous, quartz-dominated, usually friable sandstone. *Lithic grains* (rock fragments) can be recognised by their composite nature and variable colour (Figure 3.4), and they may show alteration (e.g. to chlorite). Of the *micas*, distinctive through their flaky nature (a few millimetres across), muscovite is recognised by its silvery grey colour and the less common biotite by its brown-black colour.

Some cements in arenites can be identified in the field. Apart from the acid test for calcite, many such cements are large poikilotopic crystals, several millimetres to even a centimetre across, enclosing several

Figure 3.4 *Close-up of surface of three sandstone types: (a) quartz arenite – supermature, composed of quartz grains (clear-milky, glassy) with light reflecting off crystal faces of the overgrowth cement. Shallow-marine facies. Permian, Western Australia. (b) Arkosic quartz arenite – mature, feldspar grains white from alteration to clay, some quartz overgrowths catching the light. Red pigmentation from hematite coatings of grains. Aeolian facies. Permian, NW England. (c) Lithic arenite – lithic grains of mudrock (grey-brown) and feldspar grains altered to clay (white). Fluvial facies. Carboniferous, NE England. Grains ~1 mm in diameter.*

sand grains. The cleavage fracture surfaces of such crystals are easily seen with a hand-lens, or simply with the naked eye by getting the light to reflect off the cleavage surfaces of the calcite. Quartz cement usually takes the form of overgrowths on quartz grains. Such overgrowths commonly develop crystal faces and terminations, and these too can be seen with a hand-lens, or again when the sun reflects off the crystal faces, so the grains sparkle (Figure 3.4).

3.2.1 Quartz arenites

Compositionally supermature and clean, these sandstones are typical of, but not restricted to, high-energy shallow-marine environments, and also aeolian (wind-blown) sand-seas in deserts (e.g. Figure 3.4a). Sedimentary structures are common, especially cross-stratification, on small, medium and large scales (see Section 5.3.3). Since only quartz is present, the colour of quartz arenites is commonly white or pale grey, especially those of shallow-marine environments. Aeolian quartz arenites are commonly red through the presence of finely disseminated hematite, which coats grains. Quartz and calcite cements are common.

Quartz arenites also form through leaching of a sediment, whence the unstable grains are dissolved out. *Ganister*, a type of soil occurring beneath coal seams and containing rootlets (black organic streaks), forms in this way.

3.2.2 Arkoses

Arkoses can be recognised by the high percentage of feldspar grains (more than 25%) although at outcrop these may be altered, especially to *kaolinite* (a white clay mineral) (Figure 3.4b). Many arkoses are red or pink, in part due to the presence of pink feldspars but also through hematite pigmentation. Some coarse-grained arkoses look like granites until you see the bedding. In many, grains are subangular to subrounded and sorting is moderate; a considerable amount of matrix may be present between grains. Relatively rapid erosion and deposition under a semi-arid climate produce many arkoses. Fluvial systems (alluvial fan, braided stream) are typical depositional environments for arkoses, especially if granites and granite-gneisses are exposed in the source area.

3.2.3 Litharenites

Litharenites are very variable in composition and appearance, depending largely on the types of rock fragment present. In *phyllarenites*, fragments of argillaceous sedimentary rock are dominant, and in *calclithites* limestone fragments predominate. Lithic grains of igneous and metamorphic origin are common in some litharenites. In the field it is usually sufficient to identify a rock as being a litharenite; a more precise classification would have to come from a petrographic study. The lithic grains will have variable colours, but there will likely be many feldspar grains present as well as an abundance of quartz grains (Figure 3.4c). Many litharenites are deltaic and fluvial sediments, but they can be deposited in any environment.

3.2.4 Greywackes

Greywackes are mostly hard, light- to dark-grey rocks with abundant matrix. Feldspar and lithic grains are common and often clearly identifiable with a hand-lens. Although greywackes are not environmentally restricted, many were deposited by turbidity currents in relatively deep-water basins and so show sedimentary structures typical of turbidites (sole structures, graded bedding and internal laminae; see Figures 8.3, 8.17 and 8.19). Greywackes commonly grade upwards into mudrocks.

3.2.5 Hybrid sandstones

These contain one or more components that are not detrital, such as the authigenic mineral glauconite or grains of calcite (ooids, bioclasts, etc.). The *greensands* contain granules of *glauconite* (a potassium iron aluminosilicate) in addition to a variable quantity of siliciclastic sand grains. Glauconite tends to form in marine-shelf environments starved of sediment.

Calcarenaceous sandstones contain a significant quantity (10–50%) of carbonate grains, usually skeletal fragments and/or ooids. With more than 50% carbonate grains, the rock becomes a *sandy limestone*. In *calcareous sandstones* the $CaCO_3$ is present as the cement.

For further information on the composition and mineralogy of a sandstone it is necessary to collect samples and study thin-sections made from them. *Petrofacies*, that is sandstones distinct on petrographic grounds, can be of great importance in unravelling the source of the sediment and the palaeogeography at the time. In a broad sense, the composition of a sandstone does relate to the plate-tectonic setting of the depositional basin. See sedimentary petrology textbooks for further information.

3.3 Conglomerates and Breccias

The key features that are important in the description of these sediments are the types of clast present and the texture of the rock. See Chapter 4 for textural considerations. Other terms used for these coarse siliciclastic sediments, where grains greater than 2 mm dominate, are: *rudite* (simply a coarse sedimentary rock) and *diamictite* – any poorly sorted terrigenous, generally non-calcareous, pebble-sand-mud mixture (*diamicton* is the term if uncemented). *Mixtite* has also been used. The term *megabreccia* is used for a deposit of very large blocks (see Section 5.5.1).

On the basis of clast origin, intraformational and extraformational conglomerates and breccias are distinguished. *Intraformational clasts* are pebbles derived from within the basin of deposition and many of these are fragments of mudrock or lime mudstone liberated by penecontemporaneous erosion on the seafloor/river channel, and so on, or by desiccation along a shoreline, lake margin, tidal flat, and so on, with subsequent reworking (see Figure 4.7). *Extraformational clasts* are derived from outside the basin of deposition and are thus older than the enclosing sediment (Figure 3.5; see also Figures 4.9 and 4.10).

Figure 3.5 Polymictic conglomerate, with well-rounded clasts from 5 to 10 cm of vein quartz (white pebbles), and sedimentary rocks, in a coarse sandy matrix. Tertiary, East Timor.

The variety of clasts in a conglomerate should be examined: *polymictic conglomerates* are those with several or many different types of clast (Figure 3.5); *oligomictic* (or *monomictic*) *conglomerates* are those with just one type of clast (see Figure 4.7).

The nature of the extraformational clasts in a conglomerate or breccia is important since it can give useful information on the *provenance* of the deposit, and on the rocks exposed there at the time. For a meaningful *pebble count*, several hundred should be identified but you may have to make do with less. If possible, undertake pebble counts on conglomerates from different levels in the stratigraphic succession, and on conglomerates from different parts of the region being studied. Plot the results for each locality as a histogram or pie diagram, and for the succession as a whole as a columnar diagram with different widths for the various clast types (Figure 3.6). These data could show that there were changes in the nature of rocks exposed in the source area during sedimentation and through time, as a result of uplift and erosion, or that several different areas were supplying the material.

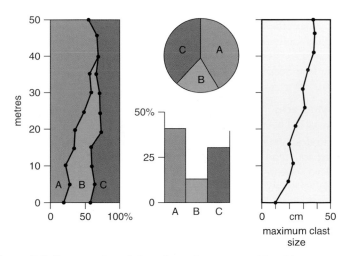

Figure 3.6 *Presentation of clast data: clast composition (three types, A, B, C) plotted stratigraphically from observations at 5-m intervals in a conglomeratic succession, and as a histogram and a pie diagram, and maximum clast size plotted stratigraphically.*

For interpretations of the depositional mechanisms of pebbly-bouldery sedimentary rocks, the texture is important: clast-supported conglomerates (also termed *orthoconglomerates*) must be distinguished from matrix-supported conglomerates (*paraconglomerates*, also called *diamictites* or *diamicton* if uncemented) (see Section 4.4). The shape, size and orientation of the pebbles should be measured (see Section 4.6), as well as the thickness and geometry of the beds and any sedimentary structures.

Conglomerates and breccias are deposited in a range of environments, but particularly glacial, alluvial fan and braided stream. Clasts in a glacial deposit, often a diamictite, may have striations and scratches. Fluvial conglomerates may be reddened, or interbedded with floodplain mudrocks with soils. Conglomerates deposited in beach and shallow-marine environments may contain marine fossils and the pebbles (especially if limestone) may have borings and encrustations of calcareous organisms. Conglomerates may also be deposited in deep water, through debris flows and high-density turbidity currents, in which case they are usually associated with mudrocks containing deep-water fossils.

Some specific types of breccia include *collapse breccia*, formed through dissolution of limestone as in a karstic breccia or evaporite (Section 4.6, see Figure 3.31), meteorite impact breccia, volcanic breccia (see Section 3.11) and tectonic breccia.

3.4 Mudrocks

Mudrocks are the most abundant of all lithologies but they are often difficult to describe in the field because of their fine grain-size. *Mudrock* is a general term for sediments composed chiefly of silt (4–62 μm) and clay (>4 μm)-sized particles. *Siltstone* and *claystone* are sediments dominated (more than 75%) by silt- and clay-grade material respectively. Claystones can be recognised by their extremely fine grain-size and usually homogeneous appearance; mudrocks containing silt or sand have a 'gritty' feel when crunched between your teeth.

Shales are characterised by the property of *fissility*, the ability to split into thin sheets generally parallel to the bedding; many shales are laminated (see Section 5.3.1). *Mudstones* are non-fissile and many have a blocky or massive texture (as in Figure 5.3). *Argillite* refers to a more indurated mudrock, whereas *slate* possesses a *cleavage* (e.g. Figure 6.7).

A *marl* is a calcareous mudrock (as in Figure 3.14). Mudrocks grade into sandstones, and terms for clay-silt-sand mixtures, and mud-sand-gravel mixtures are given in Figure 3.1.

Mudrocks are chiefly composed of clay minerals and silt-grade quartz grains; other minerals may be present. The proportion of organic matter may reach several percent and higher, and with increasing carbon content the mudrock becomes darker and eventually black in colour. A distinctive smell is produced by striking an organic-rich rock with a hammer. Hit the rock and smell the end of the hammer.

Nodules commonly develop in mudrocks, usually of calcite, dolomite, siderite or pyrite (see Section 5.5.6). Fossils are present in many mudrocks, including microfossils, which need to be extracted in the laboratory. However, macrofossils are commonly broken and compressed through compaction of the mudrock during burial.

Mudrocks can be deposited in practically any environment, particularly river floodplain, lake, low-energy shoreline, lagoon, delta, outer-marine shelf and deep-ocean basin. The sedimentological context of the mudrocks, together with the fossil content, are important in their environmental interpretation.

In the field, once the type of mudrock present has been ascertained, it can be described by the use of one or two adjectives that relate to a conspicuous or typical feature. Note the colour, degree of fissility, sedimentary structures and mineral, organic or fossil content (Table 3.2).

3.5 Limestones

Limestones, like sandstones, can only be described in a limited way in the field; the details are revealed through studies of thin-sections and peels. Three components make up the majority of limestones: carbonate grains; lime mud/micrite (micro-crystalline calcite) and cement (usually calcite spar, also fibrous calcite). The principal grains are bioclasts (skeletal grains/fossils), ooids, peloids and intraclasts. Many limestones are directly analogous to sandstones, consisting of sand-sized carbonate grains that were moved around on the seafloor, whereas others can be compared with mudrocks, being fine-grained and composed of lithified lime mud (i.e. micrite or lime mudstone). Some limestones formed in situ by the growth of carbonate skeletons, as in reef limestones (see Section 3.5.3), or through trapping and binding of sediment by microbial mats (formerly algal mats), as in *stromatolites*

Table 3.2 *Features to note and look for when examining mudrocks and examples of terms that can be used in their description.*

Mudrock feature	Possibilities and descriptive terms
Note the colour	Grey, red, black, green, variegated, mottled, marmorised
See how the mudrock breaks	Fissile (shale), non-fissile (mudstone), blocky, earthy, papery, cleaved (slate)
Look for sedimentary structures	Bedded, laminated, bioturbated, rootletted massive (apparently structureless)
Check non-clay minerals present	Quartzitic, micaceous, calcareous, gypsiferous, pyritic, sideritic, and so on
Assess the organic content	Organic-rich, bituminous, carbonaceous, organic-poor
Look for fossils	Fossiliferous, graptolitic, ostracodal, and so on

and *microbial laminites*, or through microbial precipitation, as in *thrombolites* and *tufa* (see Section 5.4.5).

The carbonate grains of modern sediments are composed of aragonite, high-Mg calcite and low-Mg calcite. Limestones are normally composed of just low-Mg calcite, with original aragonite components replaced by calcite, and the magnesium lost from original high-Mg calcite. Aragonite grains are very rarely preserved, and then only as fossils within impermeable mudrocks rather than limestones. In some limestones there are holes (molds) where originally aragonitic fossils and ooids have been dissolved out (see Figures 6.3 and 6.6). Other diagenetic changes important in limestones are dolomitisation and silicification.

Although the majority of carbonate successions in the geological record are shallow-marine in origin (supratidal to shallow subtidal), limestones are also deposited in deeper water as pelagic and turbidite beds, and in lakes. Nodular limestones, which may also be laminated and peloidal, can develop in soils and are called *calcretes* or caliches (see Section 5.5.6.2).

3.5.1 Limestone components

Skeletal grains (*bioclasts/fossils*) are the dominant constituents of many Phanerozoic limestones. The types of skeletal grain present depend on environmental factors during sedimentation (e.g. water temperature, depth, salinity) as well as on the state of invertebrate evolution and diversity at the time. The main organism groups contributing skeletal material are the molluscs (bivalves and gastropods), brachiopods, corals, echinoderms (especially the crinoids), bryozoans, calcareous algae, stromatoporoids and foraminifera. Other groups of lesser or local importance are the sponges, crustaceans (ostracods especially, also barnacles), annelids (serpulids) and cricoconarids (tentaculitids). The carbonate skeletons have different original mineralogies, and the preservation of the bioclasts in the limestone depends on this. Grains originally of low-Mg calcite, for example brachiopods, some bivalves (e.g. scallops, oysters, mussels) and serpulids, are generally very well preserved; those of high-Mg calcite originally, for example crinoids, bryozoans, calcareous red algae, rugose corals, are usually well preserved but may show some alteration. Bioclasts originally of aragonite, for example many bivalves, gastropods, scleractinian corals and green algae, are usually poorly preserved; they may be dissolved out completely (as in Figures 6.3 and 6.6) and preserved as molds or consist of coarse calcite crystals (sparite).

In the field, you should try to identify the main types of carbonate skeleton in the limestone. If these are present as macrofossils, it should be possible to identify them to the group level, and then to the genus/species level back in the lab. The carbonate skeletons may be sufficiently well preserved or abundant to allow useful palaeoecological observations to be made (see Chapter 6). One feature to check is whether the skeletal material is in its growth position, and, if it is, whether this was providing a framework for the limestone, or acting as a baffle to trap sediment, or encrusting and binding the sediment. These features are typical of reefal facies and buildups (see Section 3.5.3).

Ooids are spherical to subspherical grains, generally in the range 0.2–0.5 mm, but reaching several millimetres in diameter (Figure 3.7). Structures larger than 2 mm are referred to as *pisoids* or pisolites (Figure 3.8), and these are usually *oncoids*, of microbial origin (see Section 5.4.5). Ooids consist of concentric coatings around a nucleus,

Figure 3.7 *Close-up of oolitic grainstone with some of the millimetre-size ooids showing concentric structure. Jurassic, NE England.*

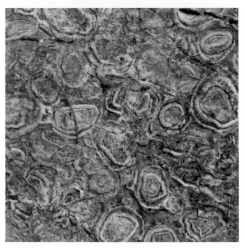

Figure 3.8 *Pisoids, 1–2 cm in diameter, which are of microbial origin so could also be called oncoids. Oncoidal bioclastic packstone. Cretaceous, Mexico.*

usually a carbonate particle or quartz grain (Figure 3.7). Most modern marine ooids are composed of aragonite, but ancient ones were generally originally calcitic in mid-Palaeozoic and Jurassic-Cretaceous times and aragonitic (now calcite or oomolds) at other times.

Peloids are subspherical to elongate grains of micrite (lime mud) generally less than 1 mm in length. They are faecal in origin or altered bioclasts.

Intraclasts are fragments of reworked carbonate sediment. Many are flakes up to several centimetres long, derived from desiccation of tidal-flat lime muds or penecontemporaneous erosion, especially by storms. The intraformational conglomerates formed in this way are sometimes called *edgewise conglomerates* or *flakestones*. *Aggregates* consist of several carbonate grains cemented together during sedimentation.

Micrite is the matrix to many bioclastic limestones and it is the main constituent of fine-grained limestones. It consists of carbonate particles mostly less than 4 μm in diameter. Much modern carbonate mud, the forerunner of micrite, is biogenic in origin, forming through the disintegration of carbonate skeletons such as calcareous algae. The origin of micrite in ancient limestones is obscure and it is often difficult to eliminate direct or indirect inorganic precipitation.

Sparite (sparry calcite, calcite spar or dolomite spar in some situations, notably Precambrian carbonates) is a clear, sometimes white and coarse, equant cement precipitated in the pore space between grains and in larger cavity structures (Figure 3.9). It is mostly a burial cement, although it can be a near-surface freshwater precipitate. *Fibrous calcite* is also a cement, coating grains and fossils and lining cavities (Figure 3.9). This is generally of marine origin and is common in reefal facies, mud mounds and stromatactis structures (see Figure 5.44 and Section 5.4.1.3).

3.5.2 Limestone types

Three schemes are currently used for the description of limestones (see Table 3.3), although the Dunham scheme is now more widely used. Common limestone types using the Folk notation are *biosparite, biomicrite, oosparite, pelsparite* and *pelmicrite*. *Biolithite* refers to a limestone that has formed through in situ growth of carbonate organisms (as in a reef) or through trapping and binding of sediment as by microbial mats to form stromatolites (see Section 5.4.5).

1. Introduction

2. Field Techniques

3. Sedimentary Rock Types

4. Sedimentary Rock Texture

Table 3.3 *Schemes for the classification of limestones. (a) On dominant grain-size. (b) On dominant constituent; prefixes can be combined if necessary, as in bio-oosparite (after R.L. Folk). (c) On dominant texture (after R.J. Dunham); see Figure 3.11 for additional Dunham terms for reef-rocks (boundstones).*

(a)

62 microns	2 mm	
calcilutite	calcarenite	calcirudite

(b)

Dominant constituent	Limestone type	
	Sparite cement	Micrite matrix
ooids	oosparite	oomicrite
peloids	pelsparite	pelmicrite
bioclasts	biosparite	biomicrite
intraclasts	intrasparite	intramicrite

In-situ growth: biolithite
Fine-grained limestone + fenestrae: dismicrite

(c)

Textural features			Limestone type
mud absent			grainstone
	grain supported		packstone
carbonate mud present		>10% grains	wackestone
	mud supported	<10% grains	mudstone
Components organically–bound during deposition:			boundstone

Figure 3.9 *Carbonate cements: a synsedimentary cavity lined with a layer of dark isopachous fibrous marine cement, followed by coarse white crystals of burial spar, containing fragments of the early cement. Mineralogy here is dolomite. Cavity is 3 cm across. Late Precambrian, California, USA.*

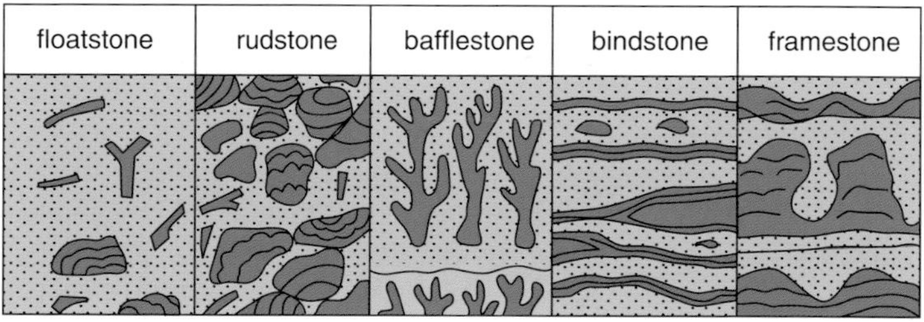

floatstone	rudstone	bafflestone	bindstone	framestone

Figure 3.10 *Sketches showing appearance of coarse limestones (most grains >2 mm), floatstone and rudstone, and three types of boundstone (reef-rock).*

Common limestones in the Dunham scheme are *grainstones*, *packstones*, *wackestones* and *mudstones;* the term *boundstone* is equivalent to biolithite. Several other terms have been introduced for reefal facies, as varieties of boundstone: *framestone, bafflestone* and *bindstone* (Figure 3.10). The term *framestone* refers to a limestone where the

1. Introduction

2. Field Techniques

3. Sedimentary Rock Types

4. Sedimentary Rock Texture

carbonate skeleton forms a framework. Robust branching corals commonly produce framestones. *Bafflestone* refers to a limestone where organisms have acted as a baffle to trap sediment; more delicate branching skeletons, including bryozoans or solitary vertically growing organisms, such as rudist bivalves and some corals, commonly form bafflestones. Platey corals, sheet-like calcareous algae and microbial mats form *bindstones*.

For coarse fossiliferous, bioclastic limestones, the terms *rudstone* – where the bioclasts (>2 mm diameter) are in contact – and *floatstone* – where the bioclasts are supported by finer sediment – are commonly used (Figure 3.10). One final type of reefal limestone is one dominated by marine cement: a *cementstone*.

The percentage of the various components present in a limestone can be estimated by reference to the charts in Figure 3.3.

In the field, careful observation of the texture and composition, often with the use of a hand-lens, can establish the type of limestone present. The main types of grain can usually be recognised with little difficulty, although in finer-grained limestones it may be impossible to distinguish between matrix and cement.

In the field the surface of a limestone is commonly weathered, especially by the action of lichens, so that the features of the rock are difficult to see. It is often necessary to examine a fresh surface, and then if you lick the rock and look with a hand-lens, you will be able to see the grains clearly.

Limestones commonly have small holes (*vugs*) through to large caverns developed within them through the effects of dissolution by rainwater or groundwater. *Speleothems* (stalactites and stalagmites) may occur within these cavities, and layers of laminated fibrous calcite (*flowstone*) often coat the limestone surface. Do not mistake modern occurrences of these features, which typically form in humid/temperate regions, for ancient examples (palaeokarsts, see Sections 5.4.1.5 and 5.4.2). And in the same way, cemented limestone breccias and calcareous soils (*calcretes*/caliches), formed in recent times, are common in limestone areas, especially in semi-arid regions; do not mistake these for ancient examples either (see Section 5.5.6.2).

The texture of the limestone (see Chapter 4) should be described, but bear in mind that the size, shape, roundness and sorting of skeletal grains in a carbonate sediment is a reflection of the size and shape of the original skeletons as well as the degree of agitation and reworking in the environment. Although practically all sedimentary structures of siliciclastics can occur in limestones, there are some that are restricted to carbonate sediments (see Section 5.4).

As noted in Chapter 2, the textural scheme of Dunham can be used directly for the graphic logging of limestones (see Figure 2.3).

3.5.3   Reef limestones

These are broadly in situ accumulations or *buildups* of carbonate material. Reef limestones have two distinctive features: a massive unbedded appearance (Figures 3.11 and 3.12; see also Figure 5.83), and a dominance of carbonate skeletons, commonly of colonial organisms, with many in their growth position. Some skeletons may have provided a framework within which and upon which other organisms grew. Cavity structures filled with internal sediment and cement are common; if the cements are fibrous they are probably marine in origin (Figure 3.9).

Reef limestones have a variety of geometries but two common forms are: *patch reef*, small and discrete structures (Figure 3.12), circular to elongate in plan; and *barrier reef*, a generally larger, usually elongate structure with lagoonal limestones behind (to landward) and reef-debris beds basinward (Figure 3.11). *Bioherm* refers to a local carbonate

Figure 3.11 *Massive reefal limestone in middle distance (cliff 50 m high) passing into well-bedded back-reef – lagoonal limestones in the foreground. Note limestone cycles in back-reef facies with internal bedding. Mid-Cretaceous, Pyrenees, Spain.*

Figure 3.12 *Small patch reef (boundstone) consisting of massive coral colonies, contrasting with bedded bioclastic gainstones below. Jurassic, NE England.*

buildup, and *biostrome* to a laterally extensive buildup, both with or without a skeletal framework. Associated with many reef limestones are beds of reef-derived debris; these may be rudstones, floatstones or grainstones/packstones. With barrier reefs and larger patch reefs especially, there is commonly a talus apron in front of or around the reef, termed fore-reef or reef-flank beds (Figure 3.11; see also Figure 5.83). These beds were often deposited on a steep to gentle slope and so show an original depositional dip (see clinoforms, Section 5.3.3.14 and Figure 5.36).

One more particular type of carbonate buildup is the *mud mound* (formerly reef knoll), consisting largely of massive lime mudstone (micrite), usually with no obvious skeletal framework organisms. Scattered skeletal debris may be present, together with cavity structures such as *stromatactis* (see Section 5.4.1.3), containing marine sediments and cements. Some mud mounds have prominent dipping flank beds; in some cases these are rich in crinoidal debris. Mud mounds typically occur within deeper-water strata; most are of Palaeozoic age, and have a microbial origin.

With all carbonate buildups, it is the massive nature that will be immediately apparent, contrasting with adjacent or overlying well-bedded limestones. Many reef-rocks have very variable textures so the terms in Figure 3.10 may apply to different areas of the same reefal limestone. In addition, the descriptive term to apply may depend on the scale of observation; a reef may be mainly a floatstone with local areas of framestone containing small patches of bindstone and cementstone, along with pockets of grainstone-to-wackestone.

Many carbonate buildups are the result of complex interactions between organisms, so verify which organisms were responsible for the buildup's construction (e.g. corals, bryozoans, stromatoporoids, rudist bivalves), which were playing a secondary but still important role of encrusting or binding the framework (e.g. calcareous algae, serpulids), and which were simply using the reef for shelter or as a source of food (e.g. brachiopods, gastropods, echinoids). There may also be evidence of bioerosion in reef-rocks; skeletons may be perforated by borings from lithophagid bivalves, serpulids or clionid sponges (see Section 5.6.2).

Within many reefs there is a clear organisation of the organisms up through the buildup. Reefs commonly initiate on bioclastic banks and shoals – mounds of skeletal debris (grainstone-rudstone), with tabular and plate-like colonial organisms growing above (a stabilisation stage generating bind-bafflestones). These create the reef (colonisation stage) and the fauna may become more varied (diversification stage) with frame-bindstones. The organisms in a reef commonly show a variety of growth forms, reflecting the energy/water depth of the environment and sedimentation rate (see Figure 3.13).

Examine the colonial fossils in the reef; sketch their shapes and see if there are any changes up through the reef.

3.5.4 Dolomites

The majority of dolomites, especially those of the Phanerozoic, have formed by replacement of limestones. This dolomitisation can take place soon after deposition, that is penecontemporaneously and notably upon high intertidal-supratidal flats in semi-arid regions, or later during shallow-burial diagenesis or deeper-burial diagenesis. For facies analysis it is important to try to decide which type of dolomite is present.

Early-formed, peritidal dolomite is typically very fine-grained and is associated with structures indicative of supratidal conditions: desiccation cracks (see Section 5.3.6), evaporites and their pseudomorphs (see Section 3.6; Figures 3.17 and 3.18), microbial laminites (see Section 5.4.5) and fenestrae (see Section 5.4.1.2). Fine-grained dolomites like these usually preserve the structures of the original sediment very well.

Later diagenetic dolomitisation can vary from local replacement of certain grains, or just the lime-mud matrix and not the grains, or just

GROWTH FORM		ENVIRONMENT	
		wave energy	sedimentation rate
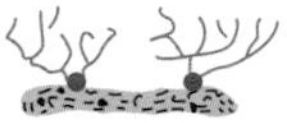	delicate branching	low	high
	thin, delicate plate-like	low	low
	columnar	moderate	high
	domal	mod-high	low
	robust, strong, branching	mod-high	moderate
	tabular	moderate	low
	encrusting	very high	low

Figure 3.13 *Different growth forms of colonial organisms (e.g. corals, stromatoporoids, rudist bivalves, calcareous algae), reflecting the local environment.*

burrows, or it may affect the whole limestone bed, the formation or just a particular facies. In some cases, just originally aragonitic and high-Mg calcite grains (bioclasts/ooids) are dolomitised and originally calcitic (low-Mg) fossils (e.g. brachiopods or oysters) are unaffected. Rhombs of dolomite may be scattered through the limestone and weather out on the surface, giving a spotted appearance. Rhombs of dolomite may be seen concentrated along stylolites. Some dolomites are very porous and have large (centimetre-size) irregular open cavities (*vugs*). Dolomite may occur in *veins* cutting through the limestone or also in *vugs* – irregular holes lined with dolomite crystals. Other minerals may be associated with these vugs, such as calcite, fluorite or galena.

Dolomite of burial origin, which commonly occurs in vugs and veins, may be of the *baroque* or *saddle* type (also called pearl spar);

this dolomite has curved crystal faces (use a hand-lens to see), maybe with steps, prominent cleavage and a pink colour from the presence of a little iron.

Many limestones are pervasively dolomitised, and then there is commonly an obliteration of the original structure of the sediment, so that fossils are poorly preserved and sedimentary structures ill-defined. With some dolomites the dolomitisation relates to tectonic structures; for example, the dolomite may occur adjacent to a fault (up which the dolomitising fluids migrated) or to major joints. The dolomitisation may be restricted to a particular stratigraphic level or certain facies, or relate to a certain stratigraphic horizon, such as occurring beneath an unconformity.

Some Precambrian dolomites show little evidence of replacement and may be of primary, or at least synsedimentary, origin. They show all the features of limestones, although stromatolites (see Section 5.4.5) are especially common. Figures 3.9 and 5.51 are Precambrian dolomites.

On the degree of dolomitisation, carbonate rocks can be divided into four categories: limestones (up to 10% dolomite), dolomitic limestone (10–50% dolomite), calcitic dolomite (50–90% dolomite) and dolomite (90–100% dolomite).

3.5.5 Dedolomites

Dedolomites are limestones formed by the replacement of dolomite. Most commonly this takes place near-surface, and in some cases it is a weathering effect. However, it may also occur in association with the dissolution of evaporites interbedded with dolomites. In some instances, unusual growth forms of calcite, such as large radiating fibrous calcite concretions ('cannonballs'), have developed within the dolomites.

3.6 Evaporites

Most *gypsum* at outcrop is very finely crystalline and occurs as white to pink nodular masses within mudrocks (which are commonly red) or as closely packed nodules with thin stringers of sediment between (*chicken-wire texture*) (Figure 3.14). Irregular and contorted layers of gypsum form the so-called *enterolithic texture*. Nodular and enterolithic textures are typical of gypsum-anhydrite precipitated in a marine sabkha (supratidal) environment, so that other peritidal sediments may be interbedded (e.g. microbial laminites/stromatolites, fenestral

Figure 3.14 *Gypsum (alabastrine) nodules (15 cm across) with associated fibrous gypsum (satin spar) veins in red marl (lacustrine) with green 'reduction' patches. Triassic, Wales.*

lime mudstones/dismicrites) or in a continental sabkha, within playa mudrocks (often red), with fluvial and aeolian sediments associated.

Beds of gypsum may also consist of large (up to a metre or more) twinned crystals (*selenite*), normally arranged vertically (Figure 3.15). This type of gypsum is typical of shallow-subaqueous precipitation.

Gypsum can be reworked by waves and storms to form *gypsarenite*, which displays current structures, and resedimented to form turbidites and slumps. Gypsum interlaminated with organic matter or calcite is typical of subaqueous (deeper water) precipitation.

Veins of *fibrous gypsum* (*satin spar*) are common in mudrocks associated with gypsum deposits (Figure 3.14). Most ancient gypsum exposed at the surface is actually secondary gypsum (alabastrine gypsum) formed by the replacement of anhydrite or primary gypsum. It can consist of centimetre-size crystals. They may show radiating patterns in the daisy gypsum variety.

Distinctive crystals of gypsum, typically several centimetres in length and colourless, are common in mudrocks that originally contained pyrite. The latter has been oxidised through near-surface weathering.

Evaporites have commonly been dissolved out of a sediment to leave a very vuggy rock. They may also be replaced by other minerals and so give rise to *pseudomorphs* of the original evaporite crystal or

Figure 3.15 *Selenite gypsum with large twinned crystals, 15 cm high. Lacustrine facies. Miocene, central Spain.*

nodule. These can be recognised in the field but confirmation may require thin-sections. Halite pseudomorphs are readily identified by their cubic shape and hopper crystal form (Figure 3.16). The lozenge, lenticular and swallow-tail shapes of gypsum crystals are distinctive (Figure 3.17). Nodules of anhydrite and gypsum can be replaced by a variety of minerals: calcite, quartz and dolomite in particular. The outside of such nodules typically has a cauliflower-like appearance (Figure 3.18). *Geodes* may form where crystals, of calcite or quartz especially, have grown inwards from the outside of the original nodule, but the whole vug has not been filled (Figure 3.18).

3.6.1 Collapse breccias

Where evaporites have dissolved away, the overlying strata have often collapsed (see Figure 3.31). If disrupted and brecciated beds occur in a section, a careful search may reveal evidence for the former presence of evaporites. Collapse breccias have angular clasts, where some fitting back together may be possible, poor sorting of clasts and little matrix. *Evaporite residues* usually consist of clayey, sandy sediments, with

Figure 3.16 *Halite pseudomorphs, up to 1 cm across, recognised by the cubic shape and depressed (hopper) crystal faces, in red playa mudstone. Lower Cambrian, central China.*

Figure 3.17 *Quartz pseudomorphs up to 10 mm across, after gypsum, recognised by the lenticular shape, in fine-grained, pale-brown dolomite. Tidal-flat facies. Late Precambrian, NW Scotland.*

dispersed clasts. Dedolomites are commonly associated with evaporite dissolution horizons.

In areas of quite severe tectonic deformation, evaporite beds are commonly the horizons along which major and minor thrusts developed. One particular type of rock that develops in this situation is a dolomitic-calcitic cellular breccia. It usually has a boxwork texture, from the dissolution of evaporite and other clasts, and it is a yellowish/buff/creamy

Figure 3.18 *Geode, 5 cm across, composed of quartz formed by replacement of an anhydrite nodule. Tidal-flat facies. Tertiary, Iran.*

colour. The French term *cargneule* (there are other spellings) is frequently used for this distinctive tectono-sedimentary rock.

3.7 Ironstones

A great variety of sedimentary rocks is included under the term ironstone and there is much variety too in the minerals present (Table 3.4). They can appear heavier than a similar-sized piece of limestone or sandstone.

Precambrian *banded iron-formations* (BIFs) are usually thick and laterally extensive deposits characterised by a fine chert-iron mineral lamination (Figure 3.19). Phanerozoic *ironstones* are mostly thin successions of limited areal extent, interdigitating with normal-marine sediments. Many such ironstones are oolitic, and ooids can be composed of hematite (red), berthierine-chamosite (green), goethite (brown) and, rarely, magnetite (black). Other common ironstones are hematitic limestones, where hematite has impregnated and replaced carbonate grains, and berthierine-chamositic, sideritic or pyritic mudrocks. All these types can be identified in the field, but later confirmation may be necessary in the laboratory.

With ironstones, interest has focused on the depositional environment and context, and so it is worth examining any fossils contained in the iron-rich beds and in adjacent strata. Check whether the fossils indicate

Table 3.4 *Main types of iron-rich sedimentary rock.*

1. Chemical iron-rich sediments

    A. Cherty iron-formation – iron minerals include hematite, magnetite, siderite, commonly in a fine lamination alternating with chert, but other varieties; mostly Precambrian

    B. Ironstone – textures similar to limestone with oolitic varieties typical; iron minerals include chamosite-berthierine, goethite, hematite; mostly Phanerozoic

2. Iron-rich mudrocks

    A. Pyritic mudrocks – pyritic nodules and laminae, often in black or bituminous shales, usually marine

    B. Sideritic mudrocks – mostly nodules in organic-rich mudrocks; often non-marine

3. Other iron-rich deposits

    A. Fe-Mn oxide-rich sediments and nodules – in oceanic facies, often associated with pillow lavas, hydrothermal activity or pelagic limestones

    B. Iron-rich laterites and soils – often developed at unconformities, also on lavas

    C. Bog-iron ores – rarely preserved in rock record

    D. Placer deposits, especially with magnetite and ilmenite

normal-marine or hyposaline (brackish-water) conditions (see Section 6.3.2). Many ironstones were formed in sediment-starved situations, in some cases in relatively deep water (outer shelf). Look at the facies above and below the ironstone to see if there is any suggestion that the ironstone accumulated at a time of maximum water depth within the succession. Otherwise treat the ironstone like any other lithology and look at its texture and sedimentary structures.

A different type of ironstone is one where grains of iron mineral, especially magnetite, are concentrated by waves and currents into laminae and beds. These *placer deposits* occur within sandstones and conglomerates deposited in fluvial and shoreline environments especially. They are easily recognised as black laminae of well-sorted grains.

The iron minerals siderite and pyrite commonly form early diagenetic *nodules* in mudrocks and other lithologies (see Section 5.5.6). Siderite

Figure 3.19 *Banded iron formation consisting of alternating layers of hematite and chert. Field of view 20 cm across. Early Proterozoic, Western Australia.*

is more typical of brackish-water muds and pyrite of marine muds. The siderite nodules are usually weathered brown on the outside, but inside are a steel-grey colour. They are common in coal measure successions, especially in seatearths (palaeosoils).

Iron and other metals are enriched in sediments associated with pillow lavas. They are usually red and brown fine-grained mudrocks. Ferro-manganese nodules and encrustations of bioclasts and lithoclasts occur in pelagic limestones and mudrocks, but they are relatively rare. They have generally formed on the seafloor in areas of strong currents. Fe-Mn impregnation of sediment and crusts may occur on hardground surfaces in pelagic limestones.

Iron-rich soils, *laterites*, are extensively developed in tropical regions and are readily recognised by their rich, deep-red to brown colour. They vary from soft and earthy to rock hard, and may show pisolitic textures. They may form a hard surface layer, that is, a *duricrust*. They do occur in the geological record, but they are not common.

3.8 Cherts

Two varieties of chert are distinguished: bedded and nodular (Figures 3.20 and 3.21).

Figure 3.20 *Bedded chert with shaley partings. Field of view 50 cm across. Basinal facies. Lower Carboniferous, S France.*

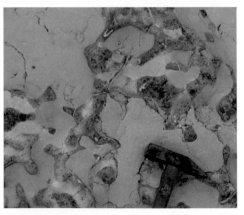

Figure 3.21 *Nodular chert in chalk. Notice the elongate and branching nature of the nodules; these have formed within a crustacean burrow system. Pelagic lime mudstone. Upper Cretaceous, NE England.*

Most *bedded cherts* are found in relatively deep-water successions and are equivalent to the radiolarian and diatom siliceous oozes of the modern ocean floors. The chert beds are usually some 3–10 cm thick, with thin (<1 cm) shale partings between (Figure 3.20). With a hand-lens on a fresh fracture surface (conchoidal fracture is typical), you can sometimes see the radiolarians in a chert sample, as minute round specks (around 0.25–0.5 mm across); a thin-section is required to check their presence. Although many beds of chert appear massive they can possess cross-lamination and graded bedding (see Sections 5.3.3 and 5.3.4) as a result of reworking on the seafloor or resedimentation into deeper water. Look at weathered surfaces for these features. Some bedded cherts are associated with pillow lavas and are part of ophiolite suites, whereas others occur in successions with no volcanic association at all.

Nodular cherts are common in limestones and some other lithologies and form by diagenetic replacement. In some cases there is a nucleus, such as a fossil (echinoid, sponge, etc.) around which replacement has proceeded; in others the nodules occur regularly spaced at particular horizons. Some chert nodules are replacements of evaporites (see Section 3.6 and Figure 3.18). *Flint* is a popular name for chert nodules occurring in Cretaceous chalks. In many cases the flint has precipitated within burrow systems (Figure 3.21), which were originally filled with sediment a little coarser than the surrounding chalk. Silica for many chert nodules is derived from dissolution of sponge spicules or siliceous plankton.

Silica can accumulate in soils to form a hard surface layer, a *silcrete*, a form of duricrust. This mostly occurs in desert regions where there are old weathered surfaces.

3.9 Phosphate Deposits (Phosphorites)

The calcium phosphate mineral in these relatively rare deposits is mostly fine-grained *collophane*, present as vertebrate bone fragments and fish scales (both of these may have a shiny black appearance), phosphatised fossils, peloids, coated grains and nodules (these are usually a dull black colour) (Figure 3.22). The nodules could be *coprolites* but they may also form by replacement of carbonate mud, grains and siliceous microfossils.

The precipitation of phosphate and the phosphatisation of sediment and fossils are commonly the result of high nutrient supply,

Figure 3.22 *Phosphate bed composed of phosphate nodules and phos-phatised fossils in pelagic lime mudstone. The dark specks are grains of glauconite (actually dark green). Sample 10 cm across. Condensed section, Mid-Cretaceous, French Alps.*

perhaps associated with upwelling, and low rates of sediment influx. Reworking is also important in the formation of many sedimentary phosphate deposits, so that they commonly form at times of increased current activity and erosion, as can be associated with a sea-level rise and transgression, for example when the quite dense phosphatic grains/clasts are concentrated into beds and lenses. *Bone beds* are usually formed in this way. Phosphatised pebbles and fossils can also be associated with hardgrounds (see Figure 5.46), which may themselves be impregnated with phosphate. Such hardgrounds occur within chalks and they may occur upon shallow-water carbonates at drowning unconformities (see Section 8.4.5). The green iron-rich mineral glauconite is present in some phosphate deposits.

3.10 Organic-Rich Deposits

Peat, brown coal (lignite), hard coal and oil shale are the main organic deposits. Bitumen and other semi-solid/solid hydrocarbons do rarely occur within sandstones and limestones, and along fault and joint planes. The organic deposits are divided into a *humic group*, those formed through in situ organic growth, chiefly in swamps, marshes and bogs, and a *sapropelic group*, where the organic matter has been transported or deposited from suspension. Most coals belong to the humic group whereas oil shales are sapropelic in origin.

The term *rank* refers to the level of organic metamorphism of a coal; a number of properties, such as carbon and volatile content, can be used to measure rank, but these require laboratory analysis.

Peat generally contains much moisture and the vegetal material may still be recognisable in a hand-specimen. It does burn of course. Peat is forming at the present time in mires, which include low-moor and high-moor locations, and in swamps and marshes around lakes and along shorelines. *High-moor peat* is dominated by mosses, especially sphagnum, and retains much water, mostly from rainfall, rather than the watertable (hence the term raised peat bog or blanket peat). There is usually little mineral matter or sediment in this peat, although the acid porewaters can leach sediment and rock below or nearby, giving hydrated iron oxide precipitates. *Low-moor (fen) peat* forms from a more diverse vegetation, including sedges, reeds and shrubs, and so is more woody. It forms close to the watertable; porewaters are less acid and there is often sediment present, mostly clays.

Some plant material is still recognisable in *soft brown coal*. In *hard brown coal* there are few plant fragments visible but the coal is still relatively soft, dull and brown. Brown coals contain much moisture when freshly dug and can be either earthy or compact. These brown coals are common in Tertiary and some older formations. *Bituminous hard coals* are black and hard with bright layers. They break into cuboidal fragments along the *cleat* (prominent joint surfaces) and make your fingers dirty. These coals are well developed in Permo-Carboniferous strata (as in Figure 5.35). *Anthracite* is bright and lustrous with conchoidal fracture. It is generally found where coal has been affected by metamorphism, or more tectonic deformation and higher heat flow.

Cannel coal and *boghead coal* are sapropelic deposits that chiefly accumulated in lakes. They are massive, fine-grained, unlaminated sediments, which possess a conchoidal fracture. Cannel coal can be carved.

Oil shales contain more than one-third inorganic material, chiefly clay but it may be carbonate. They are usually finely laminated and some can be cut with a knife into thin shavings that curl like wood shavings. They can also be set alight. They are mostly lacustrine in origin.

3.11 Volcaniclastic Deposits

This class of sedimentary rock can be difficult to study as a result of the complex processes of deposition, as well as alteration, and often limited outcrop preservation. Volcanic rocks are the result of effusive and explosive eruptions, with effusive eruptions producing lava flows and lava domes that comprise coherent and autoclastic facies, and explosive eruptions producing a wide variety of pyroclastic deposits. Thus volcanic rocks can be divided into two broad categories: (i) coherent (e.g. lava flows) and (ii) volcaniclastic. In this section we will cover the main sedimentary volcanic types and relationships (these are covered in more detail in Jerram and Petford (2011) *Igneous Rocks in the Field*, in this book series).

Coherent volcanic rocks form from cooling and solidification of molten lava and magma during effusive eruptions, which may be subaerial or submarine (even subglacial); they usually show porphyritic textures with euhedral crystals evenly distributed and in narrow size ranges, or they show aphanitic and glassy textures, with vesicles and flow foliations. These are all features of lava flows and the commonly closely associated intrusions.

Volcaniclastic rocks mostly consist (or originally consisted) of separate particles with a great range of sizes, shapes and densities, and much variation in texture and sedimentary structures too. Four main types of volcaniclastic sediment can be recognised (Table 3.5): (i) autoclastics, fragmented lava and magma formed during effusive eruptions; (ii) pyroclastics, formed during explosive eruptions; (iii) syn-eruptive resedimented volcaniclastics; and (iv) volcanogenic deposits (also called epiclastic facies). Each of these types has distinguishing features and numerous subdivisions (see Tables 3.6–3.8).

To complicate matters, *apparent volcaniclastic textures* can develop in lavas and intrusive rocks, so, for example, they can appear similar to welded ignimbrites and lithic breccias; and *apparent coherent rock textures* can develop in volcaniclastic deposits, as in welded primary pyroclastic deposits, which can resemble lava flows as a result of the coalescence of glassy pyroclasts so they are no longer distinguishable.

3.11.1 Pyroclastic material

Tephra is a general term used for unconsolidated pyroclastic material fragmented by explosive volcanic activity. Tephra consist of:

Table 3.5 *Genetic classification of volcanic deposits. A hyaloclastite is a common type of autoclastic deposit, and an ignimbrite is a common type of pyroclastic flow deposit. From McPhie et al. (1993).*

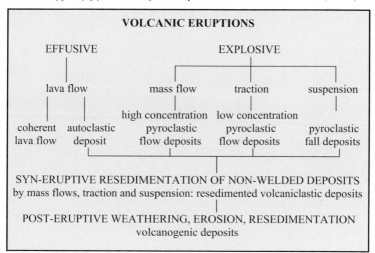

(i) pyroclasts, which include juvenile (i.e. fresh or new) fragments of lava, which range from non-vesiculated material to highly vesiculated (i.e. pumice and scoria), and these may fragment to give glass shards; (ii) crystals (phenocrysts), especially of quartz and feldspar; and (iii) lithic fragments consisting of pieces of lava from earlier eruptions (i.e. non-juvenile) and of country rock. The components of volcaniclastic deposits are given in Table 3.9.

Pyroclastic deposits are often emplaced at high temperatures. Features enabling recognition of this include: carbonised wood, pink/red coloration due to thermal oxidation, dark coloration due to finely disseminated microlites of magnetite, radial cooling joints, gas-escape structures (including fumarole pipes, vertical structures a few centimetres in width filled with coarse ash), welding together of grains and streaked-out and flattened pumice fragments.

On the basis of grain-size, the pyroclasts are divided into ash, lapilli, blocks and bombs (Table 3.10). The term *pumice* refers to light-coloured, vesicular, glassy rock of rhyolitic composition, and

Table 3.6 *Characteristic features of deposits from explosive eruptions, that is, primary pyroclastic deposits. Adapted from McPhie et al., (1993).*

Deposits from explosive magmatic and phreatomagmatic eruptions

Composed of crystals, pumice or scoria clasts, other less vesicular juvenile clasts and lithic fragments

Pumice or scoria and other juvenile clasts show porphyritic texture or are aphanitic

Abundant crystal fragments in matrix

Lithic clasts sparse to abundant

Explosive magmatic

Abundant bubble-wall glass shards in matrix

Pumice or scoria clasts usually have wispy or ragged margins and lenticular platy or blocky shapes

Accretionary lapilli occur

Welded or non-welded

Phreatomagmatic

Abundant blocky and splintery glass shards

Pumice or scoria and other juvenile clasts are typically blocky; curviplanar surfaces common

Accretionary lapilli common

Usually non-welded

Dominantly ash and fine lapilli

Deposits from phreatic eruptions

Composed of lithic pyroclasts

Hydrothermally altered clasts common; accretionary lapilli common

Small volumes ($<<1\,km^3$), limited extent ($<2\,km$ from source)

Mainly fall and surge deposits

Non-welded

scoria is used for darker pieces, still vesicular, generally of andesitic or basaltic composition. Pumice has a low density and can float on water. The vesicles in these lava fragments may be later filled with minerals such as calcite (clear), zeolite (white) or clays (green).

Table 3.7 *Characteristic features of resedimented syn-eruptive volcaniclastic deposits.*

Dominated by texturally unmodified juvenile clasts

Narrow range of clast types and composition

Sedimentation units and successions of units compositionally uniform or showing systematic upward changes

Bedforms indicate rapid deposition (mass-flow deposits common)

Table 3.8 *Characteristic features of volcanogenic sedimentary deposits (epiclastic volcanic deposits), deposited by 'normal' depositional processes.*

Mixture of volcanic and non-volcanic clasts

Volcanic clasts comprise different compositions and types

Volcanic clasts rounded

Moderate to good sorting, according to clast density

Table 3.9 *Main types of pyroclastic material (tephra) occurring in volcaniclastic deposits.*

Pumice: light coloured (low density if modern), vesiculated lava fragments, acid magma, millimetre-decimetre in size. Vesicles may contain calcite, zeolite or clay minerals

Scoria: dark-coloured equivalent of pumice

Glass shards: small grains of solidified glass, < millimetre size, formed by vesiculation and fragmentation

Glassy matrix: plastic deformation textures around hard clasts, hard and splintery, welded if hot and soft when deposited

Fiamme: compressed pumice fragments, hot and soft when deposited, millimetre-centimetre in size

Accretionary lapilli: concentrically laminated spheres of ash, 2–20 mm diameter

Lithic clasts: non-juvenile lava fragments and country rock clasts: solid when deposited, resistant to deformation, millimetre-metre in size

Phenocrysts: crystals that grew in the magma, millimetre-size

Table 3.10 *Classification of volcaniclastic grains and sediments on grain-size.*

Volcaniclastic grains		Volcaniclastic sediment terms (tephra)
bombs - ejected fluid	coarse	agglomerate
blocks - ejected solid	--- 256 mm ---	volcanic breccia
	fine	
64 mm	--------------------------------	
	coarse	
	--- 16 mm ---	
lapilli	medium	lapillistone
	--- 4 mm ---	
	fine	
2 mm	--------------------------------	
	very coarse	
	--- 1 mm ---	vitric
	coarse	/
ash	----- 0.5 mm -----	tuff --- lithic
	medium	\
	--- 0.06 mm ---	crystal
	fine	

Accretionary lapilli are small concentrically laminated spheres, 2–20 mm in diameter (like ooids) of fine volcanic ash, in some cases around a nucleus of a coarse ash grain (Figure 3.23). They have commonly formed in wet (phreatic or phreatomagmatic) eruption columns and by fall-out from steam-rich plumes.

Volcanic bombs are common in volcaniclastic successions and consist of large, usually rounded, randomly distributed 'blobs' of lava, which may depress or rupture the bedding (bomb sags) (Figure 3.24), or be deformed into strange shapes if they were still soft when they landed. Based largely on shape, spindle, bread-crust and cow-dung bombs can be distinguished.

1. Introduction

2. Field Techniques

3. Sedimentary Rock Types

4. Sedimentary Rock Texture

Figure 3.23 *Accretionary lapilli and ash in lapilli-tuff deposit. Field of view 15 cm across. Ordovician, NW England.*

Figure 3.24 *Volcanic bomb, 1 m across, with sag into underlying well-bedded ash-fall deposits. Quaternary, Santorini, Greece.*

3.11.2 Pyroclastic deposits

Three types of deposit produced by explosive volcanism are distinguished in terms of depositional process (see Tables 3.5 and 3.6).

3.11.2.1 Pyroclastic-fall deposits

Pyroclastic-fall deposits include subaerial and subaqueous (submarine or sublacustrine) fallout tephra. They are characterised by a gradual

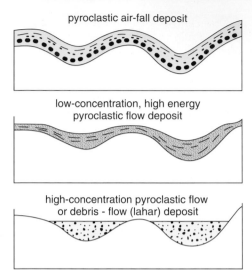

Figure 3.25 *Different geometries of pyroclastic deposits.*

decrease in both bed thickness and grain-size away from the site of eruption. Beds are typically well-sorted and normally graded (seen in Figure 3.24). These deposits can be spread over wide areas and are useful for stratigraphic correlation, forming marker beds. Pyroclastic fall deposits mantle the topography, with layers of roughly constant thickness over both hills and valleys (Figure 3.25).

Pyroclastic fall deposits are commonly reworked by currents and waves if deposited in water, or wind if subaerial, and thus may show cross- or planar lamination. Larger fragments of pumice may occur on top of the beds if they floated before being deposited. Strictly these deposits would be resedimented volcaniclastic deposits if the reworking was syn-eruptive (see Table 3.7) or volcanogenic (epiclastic) if reworking was sometime later (see Table 3.8).

3.11.2.2   Pyroclastic flow deposits

Pyroclastic flow deposits (high-particle concentration) are the product of tephra mixed with volcanic gas, and/or steam and/or water to form a density current of variable consistency; they may travel at velocities up to $100\,\text{m s}^{-1}$ One common pyroclastic flow deposit is an *ignimbrite*, produced by a violent plinian-type eruption, which generally occurs in subaerial situations, although the flows may continue into the sea or a lake.

Ignimbrites are characterised by a homogeneous appearance with little sorting of the finer ash particles and so they lack internal stratification (see Figure 3.26). Coarse lithic clasts in the bed may be normally graded (size decreasing upward), whereas large pumice clasts (which are very

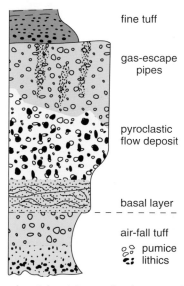

fine tuff

gas-escape pipes

pyroclastic flow deposit

basal layer

air-fall tuff

pumice

lithics

Figure 3.26 A complete 'classic' pyroclastic succession from one major eruption, which could reach several to 10 or more metres in thickness. An initial ash-fall deposit passes up into a pyroclastic flow deposit, first with a basal layer with sedimentary structures as a result of deposition from a high-velocity, low-concentration flow. This then passes up into a structureless tuff (ignimbrite) as a result of rapid deposition from a more concentrated flow, with normal grading of lithic clasts and reverse grading of pumice. The flow deposit is then overlain by fine, air-fall tuff. Fumarole pipes, formed by escaping gas and filled with coarse ash, may occur towards the top of the flow deposit.

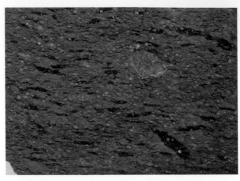

Figure 3.27 *Ignimbrite with fiamme, small crystals and fragment of lava (upper right). Field of view 15 cm. Quaternary, California, USA.*

light at the time of eruption) may show reverse grading (size increasing upwards) (Figure 3.26), or be concentrated at the top of the bed. Flattened and stretched fragments of pumice (termed *fiamme*) and glass shards indicate the soft (and hot) nature of the vesiculated and fragmented melt during transport (Figure 3.27). Many ignimbrites show welding in the central-to-lower part; here ash particles merge to form a denser, less porous rock compared with the upper and lower parts of the bed. A sub-planar foliation, termed *eutaxitic texture*, may develop here from the aligned fiamme. In its extreme, the rock is entirely glassy (vitrophyric). Lithic fragments in the deposit resist deformation and the hot plastic glassy material is deformed around them. Some ignimbrites have a columnar jointing, also indicating they were still hot on deposition (Figure 3.28). Typical thicknesses of an ignimbrite deposit are 1 to 10 m or more. The flows are topographically controlled and so the deposits fill valleys and depressions (see Figure 3.25).

3.11.2.3 Low-density pyroclastic flow deposits

Low-density pyroclastic flow deposits (sometimes referred to as surge or base-surge deposits in the literature) result from highly expanded, turbulent, gas-solid, density currents with low particle concentrations. They are characterised by well-developed unidirectional sediment bedforms (dunes) giving cross-stratification (Figure 3.29, see also

Figure 3.28 *Ignimbrite with columnar jointing (indicating hot when deposited), and eroded and weathered top surface. Overlying ashfall deposits with scattered bombs and bed thinning to left. Quaternary, Japan.*

Figure 3.29 *Pyroclastic flow deposit: basal layer with cross-bedding passing up into main ignimbrite unit, with prominent lithic clasts and crude bedding as a result of pulses in the flow. Santorini, Greece.*

Figure 5.16), pinch-and-swell features and antidune cross-bedding (see Section 5.3.3.15), since they are deposited by very fast-flowing ash-laden flows. Individual laminae are generally well sorted. These deposits tend to blanket the topography, although they do thicken into depressions (Figure 3.25). There is a complete gradation between high-particle-concentration pyroclastic flows and low-particle-concentration pyroclastic flows, and a deposit of the former may pass up into a deposit of the latter (see Figure 3.26 and Jerram and Petford, 2011).

3.11.3 Syn-eruptive resedimented volcaniclastics, and volcanogenic (epiclastic) deposits

Any volcaniclastic deposit can be reworked by normal sedimentary processes, such as wind, waves, storms, or sedimentary gravity flows, in shallow or deep water and in subaerial, sublacustrine or submarine environments. There is a spectrum of facies, from the primary pyroclastic deposits through to completely reworked and resedimented volcanogenic sediments. Two specific types of this spectrum are lahar deposits and hyaloclastites:

3.11.3.1 Lahar deposits

The flows that deposit these sediments form a continuum between syn-eruptive high-temperature flows in which hot pyroclastic material is mixed with water (such as streams or snow melt) during an eruption, and low-temperature flows representing the remobilisation by water of already deposited cool pyroclastic material, sometime later. Lahars are mudflows containing principally volcanic material, which typically deposit large volcanic fragments in a fine ashy matrix (Figure 3.30). They are characterised by a matrix-support fabric with large 'floating' clasts of volcanic material (see Section 4.4). Lahars usually have a large

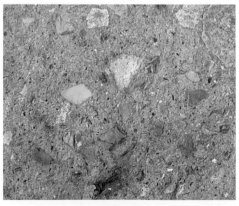

Figure 3.30 *Volcanic lahar breccia consisting of fine ash with floating angular clasts of juvenile lava, and 'old' lava (lithics). Field of view 30 cm. Quaternary, Iceland.*

range of blocks and boulders, many of which are lithic clasts rather than juvenile lava material, and an absence of any indications of very hot temperatures, compared with the hot pyroclastic flow deposits described above. Lahar deposits, by virtue of their muddy matrix, tend to be more consolidated than pyroclastic flow deposits with their ashy matrices.

3.11.3.2 Hyaloclastites

Hyaloclastites form where lava is extruded into water and the rapid chilling and quenching cause fragmentation of the lava. These *autoclastic* deposits typically consist of lava chips and flakes, a few millimetres to a few centimetres across. The chilled glassy lava is commonly altered by hydration to a yellow-green material called *palagonite*. Hyaloclastites lack any sorting or stratification close to the site of eruption but they can be reworked and resedimented to show sedimentary structures as with any other clastic sediment, then becoming resedimented volcaniclastics and volcanogenic deposits. Hyaloclastites are typical of submarine basaltic volcanism.

Where lava is intruded into or onto wet sediment it may become brecciated and intermixed to form a rock-type known as a *peperite*.

3.11.4 Studying volcaniclastic successions

When examining volcanic successions in the field it is important first to be able to recognise the textures and structures diagnostic of the emplacement-depositional processes, that is, to separate the coherent facies (lava flows and associated intrusions) from the volcaniclastic facies, autoclastic, pyroclastic, syn-eruptive resedimented volcaniclastic and volcanogenic sedimentary types; see sections above and Tables 3.6–3.8. You also need to look for features at outcrop diagnostic of the depositional setting, subaerial versus subaqueous, shallow versus deep. The original volcanic textures need to be identified and discriminated from textures attributable to reworking, alteration, deformation and/or metamorphism. Table 3.11 shows the features to look for in describing volcaniclastic deposits.

3.11.4.1 Studying younger volcaniclastic successions

It is generally easier to study young volcaniclastic deposits since you will know they are volcanic for a start (!) and they are then likely to preserve many of the original features, such as the local and regional variations in bed thickness, bed geometry and internal structures. If they

81

Table 3.11 *Descriptive terms for volcaniclastic deposits.*

Grain size: as with siliciclastic deposits (see Table 4.2) – mud/
mudstone, sand/sandstone, gravel/conglomerate or breccia

Components: crystals, crystal fragments; shards; accretionary
lapilli; lithic clasts (volcanic or non-volcanic,
monomict/polymict); vitriclasts; fiamme; pumice or scoria;
cement

Welding of grains: fusing or merging of vitric grains

Lithofacies: massive (non-bedded) or stratified (bedded) bedding:
as with siliciclastics (see Table 5.2). Graded bedding: normal,
reverse, none (see Figure 5.37). Fabric: clast or matrix
supported, sorting (see Figures 4.2 and 4.8). Jointing: blocky,
prismatic, columnar, platy

Geometry of units: following or filling topography; parallel-sided,
lenticular, tapering, channelised, lobate, and so on

Alteration: mineralogy, e.g. chloritic, sericitic, siliceous,
calcareous, hematitic, and so on. Distribution: pervasive, patchy,
disseminated, nodular, spotted

are poorly lithified, then the grain-size distributions can be studied (e.g. by sieving back in the lab) to provide useful information on the depositional processes (see textbooks). The different types of volcaniclastic deposit noted above may be identified then. Volcaniclastic deposits can be graphically logged just as other sediments, taking note of grain-size variations up through the units and section; devise symbols and legends as necessary, for example for pumice, juvenile lava clasts, lithic clasts and accretionary lapilli. There may well be lava flows interbedded with all their typical features (see below). Needless to say, working in areas of recent volcanic activity can be very dangerous.

3.11.4.2 Studying ancient volcaniclastic sediments

In more ancient strata, where erosion after deposition may have removed much of the volcanic edifice, or tectonic and metamorphic processes have obscured primary features or caused recrystallisation, it may be necessary to use a basic lithological approach to document the features, that is, noting the grain-size, composition, degree of welding, bed thickness, sedimentary structures, colour, and so on, as in any other sedimentary rock, before the depositional processes can be unravelled.

In ancient volcanic successions, care must be taken as brecciated lavas may be confused with agglomerates, and flow-banded lavas with ignimbrites. Typical features of lava flows are: (i) columnar jointing in the central part (as with some ignimbrites!); (ii) a blocky texture; (iii) brecciated basal parts and tops; (iv) vesicles concentrated towards the tops of flow units; and (v) weathered, reddened and/or rubbly upper surfaces. Ignimbrites do not usually have basal breccias.

3.11.5 Volcaniclastic successions

It is useful to document the upward changes through a volcaniclastic succession; make graphic logs and look for long-term changes in volcaniclastic deposit type.

- Are there changes in the proportion of pyroclastic fall-out versus flow deposits? Are there systematic upward bed thickness changes in the tuffs – reflecting increasing/decreasing volcanic activity?
- Are there long-term changes in the composition of the volcanic material – more acid to more basic, for example (examine the glass fragments for colour changes, the degree of vesiculation and phenocryst composition).
- Is there any evidence for compositional zonation or stratification within the magma chamber – increasing phenocryst content within a single flow deposit, for example? There may be an upward increase (or decrease) in the proportion of non-volcanic interbeds.

In some volcaniclastic successions there is a packaging of the fall-out and flow deposits, like cycles in other sedimentary strata. In a complete 'classic' pyroclastic eruption episode, the succession deposited begins with air-fall tuff, overlain by a low-density pyroclastic flow deposit, and then a high-particle-concentration pyroclastic flow deposit, before more air-fall tuff at the top of the unit, which in total may be several to many metres in thickness (see Figure 3.26).

Figure 3.31 *Collapse breccia (a deposit here of broken up dolomite beds) formed as a result of dissolution of underlying evaporite, which is now represented by a clay residue, a few cm thick, equivalent to 100 metres of gypsum in the subsurface. The residue and collapse breccia are resting on undisturbed, well-bedded dolomite forming the lower 3 metres of the cliff. Upper Permian (Zechstein) NE England.*

4

SEDIMENTARY ROCK TEXTURE

4.1 Introduction

Sediment texture is concerned with the grain-size and its distribution, morphology and surface features of grains, and the fabric of the sediment. Induration and weathering (see Section 4.7), and colour (see Section 4.8) are also considered in this chapter.

Texture is an important aspect in the description of sedimentary rocks and can be useful in interpreting the mechanisms and environments of deposition. It is also a major control on the porosity and permeability of a sediment. The texture of many sedimentary rocks can only be studied adequately with a microscope and thin-sections. With sand and silt-sized sediments you cannot do much more in the field than estimate grain-size and comment on the sorting and roundness of grains. With conglomerates and breccias, the size, shape and orientation of grains can be measured accurately in the field; in addition, surface features of pebbles and the rock's fabric can be examined quite easily. A checklist for a sediment's texture is given in Table 4.1.

4.2 Sediment Grain-Size and Sorting

The most widely accepted and used grain-size scale is that of Udden–Wentworth (Table 4.2). For more detailed work, phi units (ϕ) are used; phi is a logarithmic transformation: $\phi = -\log_2 S$, where S is grain-size in millimetres.

For sediments composed of sand-sized particles, use a hand-lens to determine the dominant grain-size class present; it is usually possible to distinguish between very coarse, coarse, medium, fine and very fine sand classes. Comparison can be made with the sand-sizes depicted in Figure 4.1. For finer-grained sediments, chew a tiny piece of the rock; silt-grade material feels gritty between the teeth compared with clay-grade material, which feels smooth.

Sedimentary Rocks in the Field: A Practical Guide, Fourth Edition   Maurice E. Tucker
© 2011 John Wiley & Sons, Ltd

Table 4.1 *Checklist for the field examination of sedimentary rock texture.*

1. Grain-size, sorting and size-grading: estimate in all lithologies: see Table 4.2 and Figures 4.1, 4.2 and 5.37. In conglomerates, measure maximum clast size and bed thickness; check for correlation

2. Morphology of constituent grains:
Shape of grains: see Figure 4.4 (important for clasts in conglomerates); look for facets on pebbles, and striations (Figure 4.11)
Roundness of grains: see Figure 4.5

3. Fabric:
(a) Look for preferred orientation of elongate clasts in conglomerates and fossils in all lithologies (see Figures 4.6, 6.7 and 6.8); measure orientations and plot rose diagram (see Chapter 7)
(b) Look for imbrication of clasts or fossils (see Figures 4.6 and 4.7)
(c) Examine matrix-grain relationships, especially in conglomerates and coarse limestones; deduce whether the sediment is matrix-supported or grain-supported (see Figure 4.8)
(d) Look for deformation of pebbles (compacted, fractured, split, pitted)

With chemical rocks such as evaporites, recrystallised limestones and dolomites, it is crystal size that is being estimated, rather than grain-size. Terms for crystal size are given in Table 4.3.

For accurate and detailed work, particularly on siliciclastic sediments, various laboratory techniques are available for grain-size analysis, including sieving of poorly cemented sedimentary rocks or modern sediments, point-counting of thin-sections of rocks and sedimentation methods (see Recommended Reading).

In the field only a rough estimate can be made of *sorting* in a sand-grade sediment. Examine the rock with a hand-lens and compare it with the sketches in Figure 4.2.

The grain-size of a sediment may fine- or coarsen-upwards through the bed to give a *graded bed*. Normal graded bedding is most common,

Table 4.2 *Terms for grain-size classes (after J.A. Udden and C.K. Wentworth) and siliciclastic rock types. For sand-silt-clay mixtures and gravel-sand-mud mixtures see Figure 3.1.*

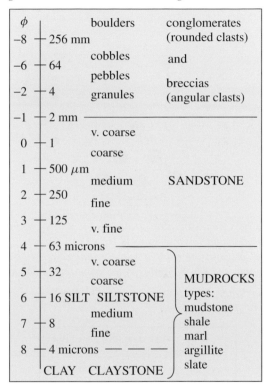

with the coarsest particles at the base, but inverse (or reverse) grading also occurs, with a coarsening up of grains. Often this is just in the lower part of a bed and then normal grading takes over. In some instances a bed may show no grain-size sorting at all. Composite graded bedding denotes a bed with several fining-upward units within it. See Section 5.3.4 for more information.

In a broad sense, the grain-size of siliciclastic sediments reflects the hydraulic energy of the environment: coarser sediments are transported

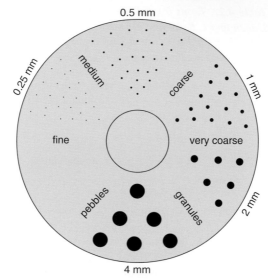

Figure 4.1 *Chart for estimating grain-size of sands: medium sand is 0.25–0.5 mm diameter, coarse sand is 0.5–1 mm diameter, and so on. Place a small piece of the rock or some grains scraped off the rock in the central circle and use a hand-lens to compare and deduce the size.*

Table 4.3 *Informal terms for describing crystalline rocks.*

	very coarsely crystalline
1.0 mm	————————————
	coarsely crystalline
0.5 mm	————————————
	medium crystalline
0.25 mm	————————————
	finely crystalline
0.125 mm	————————————
	very finely crystalline
0.063 mm	————————————
	microcrystalline
0.004 mm	————————————
	cryptocrystalline

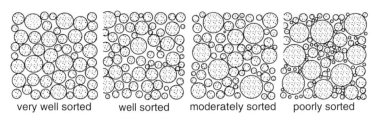

Figure 4.2 *Charts for visual estimation of sorting.*

and deposited by faster-flowing currents than those conveying finer sediments; mudrocks tend to accumulate in quieter water. The sorting of a sandstone reflects the depositional process, and this improves with increasing agitation and reworking. In contrast, the grain-size of carbonate sediments generally reflects the size of the organism skeletons and calcified hardparts that make up the sediment; these can also be affected by currents of course. Sorting terms can be applied to limestones, but bear in mind that some limestone types, oolitic and peloidal grainstones, for example, are well sorted anyway, so that the sorting terms do not necessarily reflect the depositional environment.

For grain-size and sorting of conglomerates and breccias see Section 4.6.

4.3 Grain Morphology

The morphology of grains has three aspects: *shape* (or form), determined by various ratios of the long, intermediate and short axes; *sphericity*, a measure of how closely the grain shape approaches that of a sphere; and *roundness*, concerned with the curvature of the corners of the grain.

For *shape*, four classes are recognised – spheres, discs, blades and rods, based on ratios involving the long (L), intermediate (I) and short axes (S) (Figures 4.3 and 4.4). These terms are useful for describing clast shape in conglomerates and breccias and can be applied with little difficulty in the field. The shape of pebbles is largely a reflection of the composition and any planes of weakness, such as bedding/lamination, cleavage or jointing in the rock. Rocks of a very uniform composition and structure, such as many granites, dolerites and thick sandstones, will give rise to equant/spherical pebbles; thin-bedded rocks will generally

89

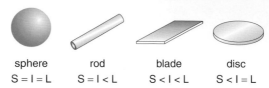

Figure 4.3 *The four common shapes of pebbles. S, I and L are the short, intermediate and long diameters, respectively.*

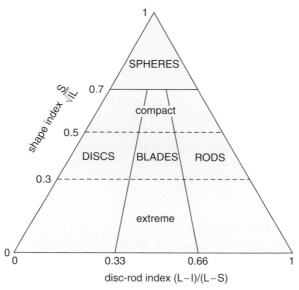

Figure 4.4 *The four classes of grain or clast shape based on the ratios of the long (L), intermediate (I) and short (S) diameters.*

form tabular and disc-shaped clasts; and highly cleaved or schistose rocks, such as slates, schists or some gneisses, will generally form bladed or rod-shaped pebbles.

Formulae are available for the calculation of sphericity and roundness (see Recommended Reading). *Roundness* is more significant than sphericity as a descriptive parameter and for most purposes the simple terms of Figure 4.5 are sufficient. These terms can be applied to

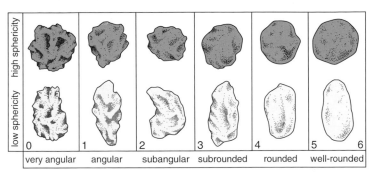

Figure 4.5 *Categories of roundness for sediment grains. For each category a grain of low and high sphericity is shown.*

grains in sandstones and to pebbles in conglomerates. In general, the roundness of grains and pebbles is a reflection of transport distance or degree of reworking.

The roundness terms are less environmentally meaningful for grains in a limestone since some, such as ooids and peloids, are well rounded to begin with. Skeletal grains in a limestone should be checked to see if they are broken or their shape has been modified by abrasion.

4.4   Sediment Fabric

Fabric refers to the mutual arrangements of grains in a sediment. It includes the *orientation* of grains and their *packing*. Fabrics may be produced during sedimentation or later during burial and through tectonic processes.

In many types of sedimentary rock a *preferred orientation* of elongate particles can be observed. This can be shown by prolate pebbles in a conglomerate or breccia, and fossils in a limestone (see, e.g., Figure 6.6), mudrock (see, e.g., Figure 6.7) or sandstone; such features are visible in the field. Many sandstones show a preferred orientation of elongate sand grains but microscopic examination is required to demonstrate this.

Preferred orientations of particles arise from interaction with the depositional medium (water, ice, wind), and can be both parallel to (the more common), and normal to, the flow direction (Figure 4.6).

SEDIMENTARY ROCKS IN THE FIELD

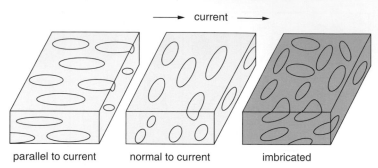

current

parallel to current   normal to current   imbricated

Figure 4.6 *Orientations of grains and pebbles: parallel to current,
normal to current, and imbricated.*

Measurement of pebble, fossil or grain orientations can thus indicate
the palaeocurrent direction (see Section 7.3.4). With pebbles it is best
to measure clasts that have a clear elongation; a length-to-width ratio of
more than 3:1 is acceptable. Preferred orientations can also be tectoni-
cally induced so if you are working in an area of moderate deformation,
also measure fold axes, cleavage and lineations. Pebbles may be rotated
into the tectonic direction. Look for pressure shadows and the development of fibrous minerals at the ends of the pebble.

Tabular and disc-shaped pebbles or fossils commonly show *imbrication*. In this fabric, they overlap each other (like a pack of cards),
dipping in an upstream direction (Figures 4.6 and 4.7). This can
be a useful texture for deducing the palaeocurrent direction (see
Section 7.3.4).

The amount of fine-grained matrix and the matrix-grain relationship
affect the *packing* and fabric of a sediment and are important in interpretations of depositional mechanism and environment. Where grains
in a sediment are in contact, the sediment is *grain-supported*; matrix
can occur between the grains, as can cement (Figures 4.7, 4.8 and 4.9).
Where the grains are not in contact, the sediment is *matrix-supported*
(Figures 4.8 and 4.10). Also look at the matrix between the large clasts
in coarser sediments; this may be well-sorted or poorly sorted (i.e.
the sediment as a whole may be bimodal or polymodal in grain-size;
see Figure 4.8).

92

Figure 4.7 *Conglomerate with a sharp-base, clast-support fabric and well-developed imbrication (elongate, flat clasts dipping down to the right) indicating transport to the left. Clasts are mudstone fragments, and so intraformational. Fluvial facies, Upper Carboniferous, NE England.*

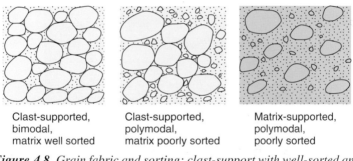

Clast-supported,
bimodal,
matrix well sorted

Clast-supported,
polymodal,
matrix poorly sorted

Matrix-supported,
polymodal,
poorly sorted

Figure 4.8 *Grain fabric and sorting: clast-support with well-sorted and poorly sorted matrix, and matrix support.*

With sandstones and limestones, a grain-support fabric with no mud generally indicates reworking by currents and/or waves/wind, or deposition from turbulent flows where suspended sediment (mud) is separated from coarser bed load. Limestones with a matrix-support fabric, such as a wackestone (see Table 3.3), mostly reflect quiet-water sedimentation. Rudstone and floatstone are coarse limestones with a grain-support and matrix-support fabric respectively (see Figures 3.10 and 4.8).

The fabric of conglomerates and breccias is discussed further in Section 4.6.

Figure 4.9 *Conglomerate with a matrix-support fabric and subangular to subrounded pebbles. Tillite (ancient glacial deposit), Late Precambrian, Scotland.*

Figure 4.10 *Polymictic conglomerate with pebble-support fabric, occurring above a massive sandstone and overlain by a sandstone with scattered pebbles and then several other thin conglomerates. Thickness of section 2 m. Braided-stream fluvial facies. Devonian, Bungle Bungles, Western Australia.*

4.5 Textural Maturity

The degree of sorting, the roundness and the matrix content in a sandstone contribute towards the textural maturity of the sediment. Texturally immature sandstones are poorly sorted with angular grains and some matrix, whereas texturally supermature sandstones are well-sorted with well-rounded grains and no matrix. Textural maturity generally increases with the amount of reworking or distance travelled; for example, aeolian and beach sandstones are typically mature to supermature, whereas fluvial sandstones are less mature. Textural maturity is usually matched by a comparable compositional maturity (see Section 3.2). It should be remembered that diagenetic processes can modify depositional texture. An estimate of the textural maturity of a sandstone can be made in the field by close examination with a hand-lens.

4.6   Texture of Conglomerates and Breccias

There is no problem with measuring the grain-sizes of these coarser sediments in the field; a ruler or tape measure can be used. With conglomerates and breccias, it is the *maximum clast size* that is usually measured. There are several ways of doing this, but one method is to take the average of the 10 largest clasts in a rectangular area of 0.5×0.5 m. It can be useful to estimate modal size as well for a conglomerate bed. Measure the long axes of 20–30 pebbles; plot a histogram and determine the size of the dominant pebbles. Maximum clast size is used as a parameter since with many rudites this is a reflection of the competency of the flow.

It is also useful to measure the *bed thickness* of conglomerates. This may vary systematically up through a succession, increasing or decreasing upwards, reflecting an advance or retreat of the source area. With some transporting and depositing processes (e.g. mudflows and stream floods) there is a positive correlation between maximum particle size and bed thickness. With braided stream conglomerates there is no such relationship.

Maximum particle size and bed thickness generally decrease down the transport path. Measurements of maximum particle size and bed thickness from conglomerates over a wide area or from a thick vertical succession may reveal systematic variations, which could be due to changes in the environment and the amount and type of sediment being supplied, and these may reflect fundamental changes involving climate or tectonics.

For the *grain-size distribution* in coarse sediments the sorting terms of Figure 4.2 can be applied, but in many cases these terms are inappropriate since the distribution is not unimodal. Many conglomerates are bimodal or polymodal in their grain-size distribution if the matrix between pebbles is considered (see Figure 4.8). It is also important to check grain-size variations through a conglomerate bed. Normal size-grading of pebbles through a bed is common but inverse/reverse grading can also occur, particularly in the basal part (see Figure 5.37 and Section 5.3.4). In some rudites, such as those deposited by debris flows, large clasts occur towards the top of the bed; these were carried there by the upward buoyancy of the flow.

The *shape and roundness* of pebbles can be described by reference to Figures 4.3 and 4.4. Taken over a large area or up a thick succession, there may be significant changes in the degree of roundness of pebbles. This can be related to the length of the transport path. With regard to shape, some pebbles of desert and glacial environments possess flat surfaces, *facets* arising either from wind abrasion (such pebbles are known as ventifacts or dreikanters) or glacial abrasion. A characteristic feature of pebbles in a glacial deposit is the presence of *striations* (Figure 4.11), although they are not always present.

The shape of pebbles may be modified during burial and through tectonic deformation. Clasts of mudrock, especially those of intraformational origin, may be folded, bent, deformed and fractured during compaction. Where there is a lot of overburden, there may be sutured contacts (stylolites) between clasts as a result of pressure dissolution (see Section 5.5.7), or one pebble may be forced into another to produce a concave pit. During more intense deformation and metamorphism, pebbles may be flattened and stretched out.

Figure 4.11 *Pebble, 12 cm across, with striae from a glacial diamictite. Permian, Western Australia.*

Attention should be given to the *fabric* of the conglomerate; in particular, check for preferred orientations of elongate clasts (if possible measure several tens, or more, of long axes) and look for *imbrication* of prolate pebbles (long axes parallel to current and dipping upstream; see Figures 4.6 and 4.7). If exposures are very good then the dip angle of the long axis relative to the bedding can be measured to give the angle of imbrication. In fluvial and other conglomerates a normal-to-current orientation is produced by rolling of pebbles, while the parallel-to-current orientation arises from a sliding of pebbles. In glacial deposits, the orientation of clasts is mostly parallel to the direction of ice movement. Glacial diamictites that have been subjected to periglacial conditions of freeze and thaw may contain split boulders.

4.6.1 Limestone breccias

In limestones some breccias are the result of in situ brecciation processes; this is the case with some karstic breccias (see Section 5.4.1.5), brecciated hardgrounds and tepees (see Sections 5.4.3 and 5.4.4), brecciated soils (calcretes; see Section 5.5.6.2) and collapse breccias formed through dissolution of intrastratal evaporites (Section 3.6, see Figure 3.31).

Examine the pebble-matrix relationship (see Section 4.4). Pebble-support fabric (Figures 4.7, 4.8 and 4.10) is typical of fluvial and beach gravels; matrix-support fabric (Figure 4.8) is typical of debris-flow deposits (debrites), which may be subaerial (as in alluvial fans or in volcanic areas; see Section 3.11 and Figure 3.30) or submarine (as in slope aprons/fans). Glacial deposits, tills and tillites, deposited directly from glacial ice, are also generally matrix-supported (Figure 4.9) and debris-flow deposits are commonly associated (the terms diamict/diamicton and diamictite are often applied to muddy gravel/conglomerate with some glacial connection; see Section 3.3).

4.7 Induration and Degree of Weathering

The induration or hardness of a sedimentary rock cannot be quantified easily. It depends on the lithology, as well as the degree of cementation, the burial history, stratigraphic age, and so on. Induration is an important concept since it does affect the degree of weathering of a rock, along with topography, climate and vegetation. A well-indurated rock in the subsurface may be rendered very friable at the surface as a result of weathering. Calcite cements in a sandstone, for example are

1. Introduction

2. Field Techniques

3. Sedimentary Rock Types

4. Sedimentary Rock Texture

easily dissolved out at the surface, as are feldspar grains and calcareous fossils. Some sandstones at surface outcrop are friable and full of holes from decalcification. On the other hand, some rocks, such as limestones, become harder on surface exposure ('case hardening'). A qualitative scheme can be used for describing induration (Table 4.4).

4.7.1 Rock exposures and outcrops

The way in which sedimentary rocks appear at outcrop can give useful information on sediment lithology, in particular the vertical changes up the succession. Mudrocks are generally less well exposed than sandstones and limestones since they are usually less well indurated and soils develop more easily upon them. Thus in cliff and mountainside exposures, sandstones and limestones tend to stand out relative to mudrocks, which weather in or are covered in vegetation. Sandstones and limestones generally give rise to steeper slopes than mudrocks. Bedding-normal joints and fractures are more common in sandstones and limestones than in mudrocks and give rise to vertical cliffs in horizontal strata. The presence of cycles in a succession, and the fining-upward or coarsening-upward of sediments in a sequence may be revealed as a result of this differential response to weathering (see, e.g., Figure 8.1).

Look at a cliff or hillside carefully; the nature of the outcrop, even if poor, the slope profile and distribution of vegetation may all give important clues to the lithologies present and upward trends and changes.

Table 4.4 *A qualitative scheme for describing the induration of a sedimentary rock.*

Unconsolidated	Loose, no cement whatsoever
Very friable	Crumbles easily between fingers
Friable	Rubbing with fingers frees numerous grains
	Gentle blow with hammer disintegrates sample
Hard	Grains can be separated from sample with penknife
	Breaks easily when hit with hammer
Very hard	Grains are difficult to separate with a penknife
	Difficult to break with hammer
Extremely hard	Sharp, hard hammer blow required
	Sample breaks across most grains

4.7.2   Weathering and alteration of sediments and rocks

The state of weathering of sediments and rocks is an important aspect of description and can give useful information on climate, present and past, and length of exposure, as well as on the degree of alteration and loss of strength for engineering purposes (see British Standards Institute, 1981). All sediments and rocks are weathered to various extents when exposed to the elements at the Earth's surface, and eventually soils with A and B zones may develop with vegetation. The weathered zone of the rocks beneath the soil is zone C. The weathering of rocks leads to discoloration, decomposition and disintegration.

Weathering features can be looked for in present-day exposures as well as in the rock record beneath unconformities. The soils above weathered zones may well be removed by subsequent erosion and so not preserved. Weathering features and soils seen at outcrop today may not be currently forming but be relict, the result of processes in the past when climate was different.

Weathering of sediments and rocks takes place through both mechanical and chemical processes with climate mostly controlling the degree of each. Mechanical weathering (temperature changes, wetting-drying) results in the opening of fractures and discontinuities and creation of new ones, at both the rock and crystal scale. Chemical weathering causes discoloration of the rock, alteration of grains, as of many silicate minerals to clays, and dissolution of grains – especially carbonates (fossils and calcite cements), and even the rock itself, leading to potholes, caverns and karst (see Section 5.4.1.5). Dissolution of limestone may lead to the residue being left behind – a quartz sand or mud, as in terra rossa soil. A weathering scale, which can be adapted to your local situation, is shown in Table 4.5 and Figure 4.12. All degrees of weathering may occur in one profile, with the A and B horizons of the soil above, or a profile may just show the lower levels as a result of erosion. Figure 4.13 shows a well-developed weathering profile.

4.8   Colour of Sedimentary Rocks

Colour can give useful information about lithology, depositional environment and diagenesis. For many purposes a simple estimate of the colour is sufficient, although it is amazing how one person's subjective impression of colour can vary from another. For detailed work, a colour chart can be used; there are several widely available including

Table 4.5 *Scale of weathering of sediment and rocks.*

Term	Description	Grade
Fresh	No visible sign of rock weathering; perhaps slight discoloration on major discontinuity surfaces	I
Slightly weathered	Discoloration indicates weathering of rock material and discontinuity surfaces. All the rock may be discoloured by weathering	II
Moderately weathered	Less than half of the rock material is decomposed or disintegrated to a soil. Fresh or discoloured rock is present either as a continuous framework or as corestones	III
Highly weathered	More than half of the rock material is decomposed or disintegrated to a soil. Fresh or discoloured rock is present either as a discontinuous framework or as corestones	IV
Completely weathered	All rock material is decomposed and/or disintegrated to soil. The original structure is still largely intact	V
Residual soil	All rock material is converted to soil. The rock structure and material fabric are destroyed. There may be a change in volume, but the soil has not been transported significantly	VI

one from the Geological Society of America based on the 'Munsell Colour System'.

It is obviously best to measure the colour of a fresh rock surface, but if different, also note the colour of the weathered surface. The latter can give an indication of the rock's composition, for example in terms of iron content.

Two factors determine the colour of many sedimentary rocks: the oxidation state of iron and the content of organic matter. Iron exists

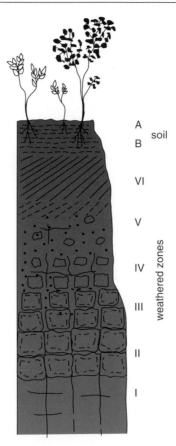

Figure 4.12 *Weathering zones of bedrock beneath a soil.*

in two oxidation states: ferric (Fe^{3+}) and ferrous (Fe^{2+}). Where ferric iron is present it is usually as the mineral hematite, and even in small concentrations of less than 1% this imparts a *red colour* to the rock. The formation of hematite requires oxidising conditions, and these are frequently present within sediments of semi-arid continental environments. Sandstones and mudrocks of these environments (deserts, playa lakes and rivers) are commonly reddened through hematite pigmentation

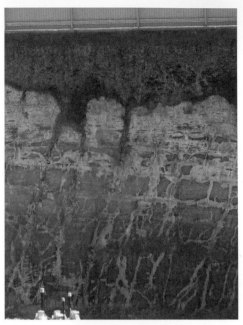

Figure 4.13 *Weathering profile upon a red calcareous mudrock. Northern Territories, Australia.*

(developed during early diagenesis) and such rocks are referred to as '*red beds*' (see, e.g., Figures 3.4b, 3.7, 4.10 and 5.3). However, red marine sedimentary rocks, for example some pelagic limestones (e.g. Ammonitic Rosso, see Figure 6.10), are also known.

Where the hydrated forms of ferric oxide, goethite or limonite are present the sediment has a *yellow-brown* or *buff colour*. In many cases, yellow-brown colours are the result of recent weathering and hydration-oxygenation of ferrous iron minerals such as pyrite or siderite, or ferroan calcite or ferroan dolomite.

Where reducing conditions prevailed within a sediment, the iron is present in a ferrous state and generally contained in clay minerals; ferrous iron imparts a *green colour* to the rock. Green colours can develop through reduction of an originally red sediment, and vice versa (see

Figure 3.14). With red- and green-coloured deposits see if one colour, usually the green, is restricted to, say, coarser horizons or is concentrated along joint and fault planes; this would indicate later formation through the passage of reducing waters through the more permeable layers or conduits.

Organic matter in a sedimentary rock gives rise to *grey colours*, and with increasing organic content to a *black colour*. Organic-rich sediments generally form in anoxic conditions. Finely disseminated pyrite also gives rise to a dark grey or black colour. *Black pebbles*, which are reworked out of soil horizons and may be the result of forest fires, are commonly associated with unconformities and exposure horizons.

Other colours such as olive and yellow can result from a mixing of the colour components. Some minerals have a particular colour and if present in abundance they can impart a strong colour to the rock; for example glauconite and berthierine-chamosite give rise to green-coloured sediments. Anhydrite, although not normally present at outcrop, may be a pale blue colour.

Some sediments, especially mudrocks, marls and fine-grained limestones, may be *mottled*, with subtle variations in grey, green, brown, yellow, pink or red colours. This may be due to bioturbation and the differential colouring of burrows and non-bioturbated sediment (*burrow mottling*), generating an *ichnofabric* (see Section 5.6.1), or it may be due to pedogenic processes: water moving through a soil causing an irregular distribution of iron oxides-hydroxides and/or carbonate, and/or the effect of roots and rhizoturbation (see Section 5.5.6.2). The term *marmorisation* has been applied to this process. Colour mottling

Table 4.6 *The colour of sedimentary rocks and probable cause.*

Colour	Probable cause
Red	hematite
Yellow/brown	hydrated iron oxide/hydroxide
Green	glauconite, chlorite
Grey	some organic matter
Black	much organic matter
Mottled	partly leached
White/no colour	leached

is common in lacustrine and floodplain muds and marls (see, e.g., Figure 5.3), especially sediments of palustrine facies (lake sediments strongly affected by pedogenesis).

Many sedimentary rocks show curious colour patterns (see Figure 4.14) that are similar to those produced in chromatography or loosely referred to as *Liesegang rings*: swirling, curved and cross-cutting patterns that are oblique to the bedding. The colours are usually shades of yellow and brown, even red, from variations in the contents of iron oxides and hydroxides. These may form at any time after deposition, although often related to weathering, and relate to the passage and diffusion of porewater through the sediment and precipitation or dissolution of minerals.

The common colours of sedimentary rocks and their cause are shown in Table 4.6.

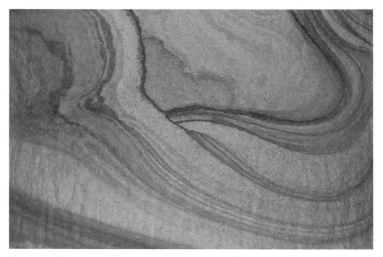

Figure 4.14 Iron-rich/iron-poor patterns (liesegang rings) in a fluvial sandstone resulting from pauses in the movement of groundwater through the sediment. The patterns usually have nothing to do with deposition or primary sedimentary structures. Carboniferous, Durham Castle, NE England.

5

SEDIMENTARY STRUCTURES AND GEOMETRY OF SEDIMENTARY DEPOSITS

5.1 Introduction

Sedimentary structures are an important attribute of sedimentary rocks. They occur on the upper and lower surfaces of beds as well as within beds. They can be used to deduce the processes and conditions of deposition, the directions of the currents that deposited the sediments (see Chapter 7), and in areas of folded rocks, the way-up of the strata (see Section 2.9). An index of sedimentary structures as presented in this book is given in Table 5.1.

Sedimentary structures are very diverse and many can occur in almost any lithology. Sedimentary structures develop through physical and/or chemical processes before, during and after deposition, and through biogenic processes. It is convenient to recognise five categories of sedimentary structure: erosional (see Section 5.2); depositional – all sediment types (see Section 5.3); depositional – especially in limestones (see Section 5.4); post-depositional/diagenetic (see Section 5.5) and biogenic (see Section 5.6).

The geometry of sedimentary deposits is an important feature at all scales and is discussed in Section 5.7, as are the relationships between sedimentary units.

5.2 Erosional Structures

The common structures of this group are the flute, groove and tool marks that occur on the undersurfaces (soles) of beds, scour structures in general, and channels.

Sedimentary Rocks in the Field: A Practical Guide, Fourth Edition   Maurice E. Tucker
© 2011 John Wiley & Sons, Ltd

Table 5.1   *Index of sedimentary structures: main types and location of description and/or figures in this book.*

Bedding surface structures

Ripples: look at symmetry/asymmetry and crest shape; current, wave or wind ripples? Section 5.3.2, Figures 5.9–5.12

Shrinkage cracks: desiccation or syneresis cracks: Section 5.3.6, Figures 5.39 and 5.40

Parting lineation (primary current lineation) Section 5.3.1.3, Figure 5.7

Rainspot impressions: Section 5.3.7

Tracks and trails: crawling, walking, grazing, resting structures. Section 5.6.2, Tables 5.7 and 5.9, Figures 5.71 and 5.72

Bedding undersurface (sole) structures

Flute casts: triangular, asymmetric structures. Section 5.2.1, Figure 5.1

Groove casts: continuous/discontinuous ridges. Section 5.2.2, Figure 5.2

Tool marks: Section 5.2.3

Load casts: bulbous structures. Section 5.5.5, Figures 5.56 and 5.57

Scours and channels: small and large-scale. Sections 5.2.4 and 5.2.5, and Figures 5.3, 5.35, 5.38 and 8.13

Shrinkage cracks: Section 5.3.6

Internal sedimentary structures

Bedding and lamination: Section 5.3.1, Table 5.2, Figures 5.4–5.8

Graded bedding: normal and reverse. Section 5.3.4, Figures 3.23, 4.7 and 5.37

Cross-stratification: many types; see Table 5.3 and Section 5.3.3

Massive bedding: is it really massive? Section 5.3.5, Figure 5.38

Slumps and slumped bedding: Section 5.5.1, Figures 5.52 and 5.53

Deformed bedding: various specific types. Section 5.5.2, Figure 5.54

Sandstone dykes: Section 5.5.3

Dish structures: concave-up laminae, pillars between. Section 5.5.4, Figure 5.55

(*continued overleaf*)

Table 5.1 *(continued)*

Nodules: Section 5.5.6, Table 5.4, Figures 5.58–5.60
Stylolites: sutured planes. Section 5.5.7, Figures 5.64–5.66
Joints and fractures: Section 5.5.8
Burrows: feeding and dwelling biogenic structures. Section 5.6,
    Figures 5.70, 5.73–5.78

**Structures restricted to or predominant in limestones (and
    dolomites)**
Cavity structures: usually filled with calcite cement. Section 5.4.1
Geopetal structures: Figures 3.9, 5.41 and 5.42
Birdseyes: Figure 5.43; laminoid fenestrae, usually in microbial
    limestones
Stromatactis: cavity with flat base and irregular roof, Figure 5.44
Palaeokarstic surfaces: irregular, potholed surface. Section 5.4.2,
    Figure 5.45
Hardgrounds: recognised by encrusted and bored surfaces. Section
    5.4.3, Figures 5.46 and 5.47
Tepees: pseudoanticlinal structures. Section 5.4.4
Stromatolites: planar-laminated sediments, columns, domes,
    Section 5.4.5, Figures 5.48–5.51
Collapse structures: rock dissolution and collapse of overlying
    strata. Sections 3.6 and 5.4.1.5

5.2.1 Flute casts

Flute casts are readily identifiable from their shape (Figure 5.1). In plan, on the bedding undersurface, they are elongate to triangular ('heel-shaped') with either a rounded or pointed upstream end; they flare in a downstream direction. In section they are asymmetric, with the deeper part at the upstream end. Flute marks vary in length from several to tens of centimetres. Flutes form through erosion of a muddy sediment surface by eddies in a passing turbulent current and then the marks are filled with sediment as the flow decelerates. Flute casts are typical of sandstone *turbidites*. They also occur on the underside of fluvial sandstones, such as those deposited during crevassing, when a river current moves across a floodplain, and on the bases of

Figure 5.1 *Flute marks on undersurface of a siliciclastic turbidite bed. Current flow from lower right to upper left. Field of view about 1 m across. Deepwater facies. Cambrian, southern France.*

sandstones and limestones deposited by storm currents ('*tempestites*'; see Section 5.3.3.9). However, those on the bases of turbidites tend to be more uniform in size, regular in shape and evenly spaced and organised; they often cover larger areas of the bed undersurface.

Flute marks are reliable indicators of palaeocurrent direction; their orientation should be measured (see Chapter 7).

5.2.2 Groove casts

Groove casts are elongate ridges on bed undersurfaces, ranging in width from a few millimetres to several tens of centimetres (Figure 5.2). They may fade out laterally, after several metres, or persist across the exposure. Groove casts on a bed undersurface may be parallel to each other or they may show a variation in trend, up to several tens of degrees or more. Groove casts form through the filling of grooves, cut chiefly by objects (lumps of mud or wood, etc.) dragged along by a current. Groove casts are common on the undersurfaces of turbidites. Similar structures, albeit usually less regular or persistent, can occur on the soles of some fluvial sandstones and storm-deposited sandstones/limestones; the term *gutter cast* is often applied to these. Groove/gutter casts indicate the trend of the current and their orientation should be measured (see Chapter 7).

Figure 5.2 *Groove marks on undersurface of siliciclastic turbidite. Hammer 30 cm long. Deep-water facies. Silurian, southern Scotland.*

5.2.3 Tool marks

These form when objects being carried by a current come into contact with the sediment surface. The marks are referred to as prod, roll, brush, bounce and skip marks, as appropriate, or simply as tool marks. An impression left by an object may be repeated several times, if it was saltating (bouncing along). Objects making the marks are commonly mud clasts, pebbles, fossils and plant debris. Once made, the impression of a tool may be eroded and elongated parallel to the current direction. As with flutes and grooves, casts are formed when sediment fills the tool mark and so they are usually seen on the soles of sandstone and limestone beds. They are particularly common on the bases of turbidite beds.

5.2.4 Scour marks and scoured surfaces

These are structures formed by current erosion. The term scour mark would be used for a small-scale erosional structure, generally less than a metre across, cutting down several centimetres, and occurring on the base of, or within, a bed. In plan they are usually elongate in the current direction. With increasing size, scours grade into channels (see Section 5.2.5). Typical features of scoured surfaces are the cutting out of underlying sediments, the truncation of underlying laminae and the presence of coarser sediment overlying the scoured surface (see Figure 5.55). The scoured surfaces are usually sharp, irregular with some relief, but they can be smooth.

Scour marks and scoured surfaces are restricted neither to lithology nor environment but occur wherever currents are sufficiently strong to erode into underlying sediment. They usually form during a single erosional event.

5.2.5 Channels

Channels are larger-scale structures, metres to kilometres across, that are generally sites of sediment transport for relatively long periods of time. Many channels are concave-up in cross-section (Figure 5.3) and their fills may form elongate (shoestring) sediment bodies when mapped out. As with scours, channels can be recognised by their cross-cutting

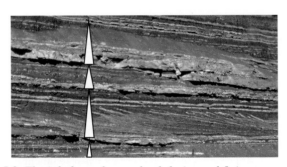

Figure 5.3 *Fluvial channels, overbank facies and fining-upward cycles (indicated by triangles) in a floodplain succession. Lower channel shows lateral accretion, i.e. meandering, left to right, and is mostly filled with mudrock. Upper sand-filled channels show cross-bedding. The channel deposits fine up into floodplain mudrocks of red and green colours. The pink-white nodular layers are gypsum; one is cut out by the lowest channel. Triassic, Arizona, USA.*

110

relationship to underlying sediments (Figures 5.3, 5.35 and 5.38). Channels are usually filled with sediment coarser than that below or adjacent, and may be a basal conglomeratic layer (a lag deposit). Cross-bedded sandstones fill many channels.

Some large channels may not be immediately apparent in the field; thus, view quarry faces and cliffs from a distance and take careful note of the lateral persistence of sedimentary units. Sediments within channels may onlap the sides of the channel (see Section 5.7). Channel-fill sediments commonly show upward changes in grain-size (usually fining-up) or facies, for example from fluvial to estuarine to marine as an *incised valley* is gradually filled through a relative sea-level rise.

Channels are present in the sediments of many different environments, including fluvial, deltaic, shallow subtidal-intertidal and submarine fan. With fluvial, deltaic and tidal channels, check for evidence of lateral accretion (see Section 5.3.3.12), that is, low-angle dipping surfaces, which would indicate lateral migration (meandering) of the channel (Figures 5.3 and 5.35). Try to measure the orientation of the channel structure, for example from the largest scale cross-bedding; this usually indicates the trend of the palaeoslope, important in palaeogeographical reconstructions.

5.3 Depositional Structures

In this group are the familiar structures of bedding, lamination, cross-stratification, ripples and mudcracks. Depositional structures occur on the upper surface of beds and within them. In limestones additional structures that may be present include various types of cavity, features produced by synsedimentary cementation (hardgrounds and tepees) and subaerial dissolution (palaeokarstic surfaces, cavities and breccias), and stromatolites-thrombolites (see Section 5.4).

5.3.1 Bedding and lamination

Bedding and lamination define stratification. Bedding is thicker than 1 cm whereas lamination is thinner than 1 cm. Bedding is composed of beds; lamination is composed of laminae. Parallel (also called planar or horizontal) lamination is a common internal structure of beds. Descriptive terms for bed and laminar thickness are given in Table 5.2. Beds also vary in their shape and definition so that planar, wavy and curved types are recognised, and these may be parallel

Table 5.2 *Terminology of bed thickness.*

	Very thickly bedded
1 metre	
	Thickly bedded
0.3 m	
	Medium bedded
0.1 m	
	Thinly bedded
0.03 m	
	Very thinly bedded
10 mm	
	Thickly laminated
3 mm	
	Thinly laminated

to each other, non-parallel or discontinuous. Figure 5.4 illustrates these variations.

5.3.1.1 Bedding

This is produced by changes in the pattern of sedimentation; it may be defined by changes in sediment grain-size, colour or mineralogy-composition. *Bed boundaries* may be sharp, smooth or irregular, or gradational. There are commonly thin shale seams or mud partings at the contacts between sandstone beds and limestone beds. Bedding-plane surfaces may be smooth, undulating, rippled, sutured, and so on, and represent long or short breaks in sedimentation. The range of bed contacts and features is shown in Figure 5.5.

- Check for evidence of erosion (scour) at bed boundaries; i.e. is there a sharp erosional contact with coarser sediment above?
- Look for grain-size or composition changes towards the top of the bed below the bedding plane; is there any evidence for shallowing-up towards the surface?

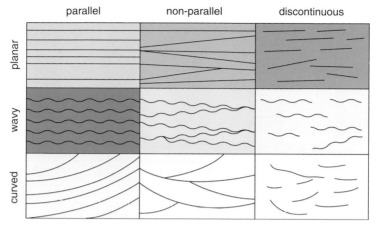

parallel · non-parallel · discontinuous

planar

wavy

curved

Figure 5.4 *Different types of bedding or lamination.*

- Is the bedding plane an exposure horizon? Look for mudcracks, rootlets, karstic features.
- Examine bedding-plane surfaces for such structures as ripples, parting lineation, mudcracks and rootlets.
- Look at bed undersurfaces for erosional structures such as the casts of flutes, grooves and tool marks.
- Examine bed cross-sections for internal sedimentary structures such as cross-stratification and graded bedding. One important point to decide is whether the bed was deposited in a single event (e.g. by a turbidity or storm current) or over a much longer period of time (years to hundreds of years).

With *limestones*, bedding planes can be palaeokarstic surfaces, denoting emergence, or hardground surfaces produced through synsedimentary submarine cementation at a time of reduced or negligible sedimentation (see Sections 5.4.2 and 5.4.3). Be aware, however, that bedding planes, particularly those in limestones and dolomites, can be enhanced, modified or even produced by pressure dissolution during burial to give sutured stylolitic surfaces or more smooth undulating dissolution seams (see Section 5.5.7).

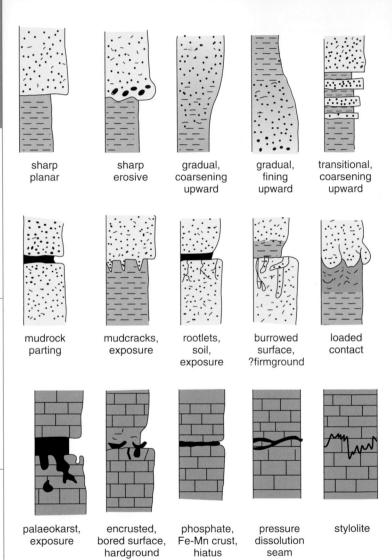

sharp
planar

sharp
erosive

gradual,
coarsening
upward

gradual,
fining
upward

transitional,
coarsening
upward

mudrock
parting

mudcracks,
exposure

rootlets,
soil,
exposure

burrowed
surface,
?firmground

loaded
contact

palaeokarst,
exposure

encrusted,
bored surface,
hardground

phosphate,
Fe-Mn crust,
hiatus

pressure
dissolution
seam

stylolite

Figure 5.5 *Bedding planes and bed contacts: the range of possibilities.*

- Look for the lateral continuity of bedding surfaces; if depositional, they should be very persistent; if diagenetic (pressure dissolution) they may fade out or cut up/down through the bed.

Bed boundaries can be deformed by soft-sediment compaction and loading; the contacts between sandstones and underlying mudrocks are commonly affected in this way (see Section 5.5.5). Tectonic movements, bedding-plane slip, for example, and the formation of cleavage, can also modify bed junctions.

5.3.1.2 Bed thickness

This is an important and useful parameter to measure. With some current-deposited sediments, turbidites and storm beds ('tempestites'), for example, bed thickness decreases in a down-current direction. In vertical succession, there may be a systematic upward decrease or increase in bed thickness, reflecting a gradual change in one of the factors controlling deposition (e.g. increasing/decreasing distance from the source area, or increasing/decreasing amount of uplift in the source area affecting sediment supply). Alternatively, the beds may be arranged into small repeated units of bed thickness decreasing/increasing upwards (see Section 8.4.3). In submarine fan successions, for example, units a few metres thick of turbidite beds may show an upward increase in bed thickness, reflecting the growth of suprafan lobes (see Figure 8.3). With some conglomerates, there is a relationship between bed thickness and sediment grain-size (Section 4.6).

5.3.1.3 Parallel/horizontal lamination and flat-bedding

Thin layering, in sandstones, limestones and mudrocks, is defined by changes in grain-size, mineralogy, composition and/or colour and can be produced in several ways. *Flat-bedding* in sandstones and limestones can be formed through deposition from both strong currents, referred to as *upper plane-bed phase lamination*, or from weak currents in *lower plane-bed phase lamination*. Lamination in mudrocks is formed through deposition from suspension and low-density turbidity currents, and from mineral precipitation.

Upper plane-bed phase lamination predominantly occurs in sandstones and limestones and forms through subaqueous deposition at high flow velocities in the upper flow regime (see sedimentology textbooks). Laminae are several millimetres thick and are made visible by subtle

Figure 5.6 *Flat bedding formed from fast-flowing currents in the upper plane-bed phase, overlain by cross-bedding. Shoreface quartzitic lime-stone with fossil bivalves. Field of view 50 cm. Pleistocene, Western Australia.*

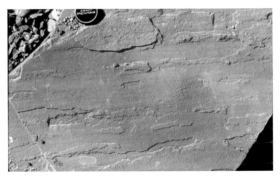

Figure 5.7 *Parting lineation (or primary current lineation), trending from left to right of photograph, in flat-bedded turbidite sandstone. Field of view 40 cm. Basinal facies. Cretaceous, California.*

grain-size changes (see, e.g., Figure 5.6). This type of parallel lamina-tion is characterised by the presence of a *parting lineation*, also called *primary current lineation*, on the bedding surfaces (Figure 5.7). Such surfaces, when seen in the right 'light', have a visible fabric or streak-iness consisting of very low ridges only several grain diameters high. This lineation is produced by turbulent eddies close to the sediment surface. Parting lineation is formed parallel to the flow direction, so that its orientation will indicate the trend of the palaeocurrent.

SEDIMENTARY STRUCTURES

Lower plane-bed phase lamination lacks parting lineation and occurs in sediments with a grain-size coarser than 0.6 mm (i.e. coarse sand grade). It forms through the movement of sediment as bed load, by traction currents at low flow velocity in the lower flow regime. It commonly occurs in coarse sandstones and limestones.

Lamination formed largely by deposition from suspension or low-density turbidity currents occurs in a wide range of fine-grained lithologies, but especially mudrocks, fine-grained sandstones and limestones. Laminae are typically a few millimetres thick and normally graded (see Section 5.3.4) if deposited from suspension currents, as is the case with varved and rhythmically laminated sediments of glacial and non-glacial lakes (Figure 5.8).

Lamination can also be produced by the periodic precipitation of minerals such as calcite, halite or gypsum/anhydrite, and from the blooming

Figure 5.8 *A sub-millimetre-scale rhythmic lamination, possibly of annual/seasonal origin, of alternating clay-rich and carbonate-rich laminae. Note the bundling of the laminae into units about 1.5 cm thick. Thicker units are distal turbidites. Height of section 20 cm. Permian lower-slope-basinal lime mudstone, NE England.*

of plankton in surface waters with subsequent deposition of organic matter. Many finely laminated fine-grained sediments are deposited in protected environments such as lagoons and lakes and in relatively deep-water marine basins below the wave-base.

In the field, use a hand-lens to see the cause of the lamination:

- Is it a fine intercalation of different lithologies (e.g. claystone/siltstone or claystone/limestone) or a grain-size change (e.g. graded silt to clay laminae) or both?
- If in a sandstone, split the rock and look for parting lineation on bedding surfaces.
- If in a limestone, check that the lamination has been formed by the physical movement of grains and that it is not microbial (i.e. stromatolitic) in origin (see Section 5.4.5).
- Measure the thickness of laminae and the thickness of the parallel-laminated/flat-bedded unit.
- Look for any bundling of the laminae (as in Figure 5.8), which may suggest a longer-term control on deposition. For example, if the laminae were annual/seasonal, there may be evidence for a longer-term climatic control affecting lamina thickness through time (e.g. sun-spots).

5.3.2 Ripples, dunes and sand-waves

These are bedforms developed chiefly in sand-sized sediments, limestones or sandstones, and even cherts, gypsum (gypsarenite) and ironstones. Ripples are very common and occur on bedding surfaces, but the larger-scale dunes and sand-waves are rarely preserved intact as the bedforms. The migration of ripples, dunes and sand-waves under conditions of net sedimentation gives rise to various types of cross-stratification (see Section 5.3.3), which is one of the most common internal depositional structures in sandstones, limestones and other sedimentary rock-types. Both wind and water can move sediment to produce these structures.

5.3.2.1 Wave-formed ripples

These are formed by the action of waves on non-cohesive sediment, especially the medium silt to coarse sand grades, and they are typically symmetrical in shape. Asymmetrical varieties do occur, formed when one direction of wave motion is stronger than the other, and they may be

118

Figure 5.9 *Wave-formed ripples, with crest bifurcation and small ripples in the troughs of the larger ones (wavelength 10 cm) formed at low water. Shoreline sandstone. Meso-Proterozoic, Western Australia.*

H ← L →	ripple index = L/H
wind ripples L 2.5–25 cm H 0.5–1.0 cm	mostly 10–70
wave ripples L 0.9–200 cm H 0.3–25 cm	4–13 mostly 6–7
current ripples L < 60 cm H < 6 cm	>5 mostly 8–15

Figure 5.10 *The ranges of wavelength (L), height (H) and ripple index for wind, wave-formed and current ripples.*

difficult to distinguish from straight-crested current ripples. The crests of wave-formed ripples are generally straight, and the bifurcation of crests is common (Figure 5.9); the crests may rejoin to enclose small depressions (called tadpole nests!). In profile, the troughs tend to be more rounded than the crests, which can be pointed or flattened. The ripple index (Figure 5.10) of wave-formed ripples is generally around 6 or 7. Wavelength is controlled by sediment grain-size and water depth, larger ripples occurring in coarser sediment and deeper water.

Wave-formed ripples can be affected by changes in water depth to produce modified ripples, ripples with flat crests or double crests, for example. If there is a change in the direction of water movement over an area of ripples (current or wave-formed), then a secondary set of ripples can develop, producing *interference ripples*, or small ripples may form within the troughs of larger ripples (ladderback ripples). Modified and interference ripples are typical of tidal-flat and shallow-water deposits.

5.3.2.2 Current ripples, dunes and sand-waves

Current ripples are produced by unidirectional currents so they are asymmetric with a steep lee-side (downstream) and gentle stoss-side (upstream) (Figure 5.11). On the basis of shape, three types of current ripple are common: straight-crested, sinuous or undulatory, and linguoid ripples (Figure 5.12). Lunate ripples do occur but are rare. With increasing flow velocity of the current, straight-crested ripples pass into linguoid ripples via the transitional sinuous ripples. The ripple index (Figure 5.10) of current ripples is generally between 8 and 15. Current ripples do not form in sediment coarser than 0.6 mm diameter (coarse sand). Current ripples can develop in almost any environment: river, delta, shoreline, offshore shelf and deep sea.

Subaqueous dunes (also called megaripples) and sand-waves (sand-bars) are larger-scale structures of similar shape to ripples. Although rarely preserved intact, the cross-bedding produced by their migration is a very common structure indeed (see Section 5.3.3.2). Subaqueous

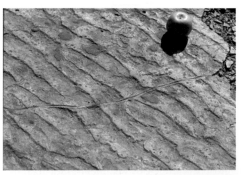

Figure 5.11 *Current ripples. These asymmetric straight-crested ripples are transitional to linguoid in places. An animal trail crosses the ripples, and the round holes are probably annelid burrows. Current flow from right to left. Deltaic facies. Mid-Carboniferous, NE England.*

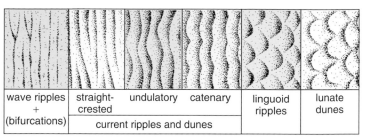

wave ripples + (bifurcations)	straight-crested	undulatory	catenary	linguoid ripples	lunate dunes
	current ripples and dunes				

Figure 5.12 *Crest plans of wave ripples, current ripples and dunes. Linguoid dunes and lunate current ripples are rare. Stoss sides (less steep, upstream facing) are stippled, i.e. current flowing left to right. Dunes are larger-scale bedforms than ripples (see text).*

dunes are generally a few metres to more than 10 m in wavelength and up to 0.5 m high. They can be seen in modern rivers and estuaries. Dune shape varies from straight-crested to sinuous to lunate with increasing flow velocity. Ripples commonly occur on the stoss-sides and in the troughs of dunes. Sand-waves are larger than dunes, being hundreds of metres in wavelength and width, and up to several metres in height. Many are linguoid in shape. Sand-waves can be seen in large rivers, and similar structures occur on shallow-marine shelves. In rivers, sand-waves form at lower flow velocities and at shallower depths than dunes.

5.3.2.3 Wind ripples and dunes

These are asymmetric structures like current ripples. Wind ripples typically have long, straight, parallel crests with bifurcations like wave-formed ripples. The ripple index is high (Figure 5.10), that is, the ripples are quite flat. Wind ripples are rarely preserved. The dunes produced by wind action are also rarely preserved themselves, because of their size, but the cross-stratification produced by their migration is a feature of ancient desert sandstones (see Section 5.3.3.11). The two common aeolian dune types are barchans (lunate structures) and seifs (elongate sand ridges), and these may occur within or upon larger areas of wind-blown sand. The existence of very large seif dunes and draas may be revealed by mapping the distribution and thickness of aeolian sandstones over a wide area.

5.3.3 Cross-stratification

Cross-stratification is an internal sedimentary structure of many sand-grade, and coarser, sedimentary rocks and consists of a stratification at an angle to the principal bedding direction. Terms formerly used, such as current bedding, false or festoon bedding, are best avoided. Much cross-stratification is formed as a result of deposition during the migration of ripples, dunes and sand-waves. However, cross-stratification in sand-grade sediments can also be formed through the filling of erosional hollows and scours, the growth of small deltas (as into a lake or lagoon), the development of antidunes and hummocks, the lateral migration of point bars in a channel and deposition on a beach foreshore. Large-scale cross-bedding is typical of aeolian sandstones. Cross-bedding can also form in conglomerates, notably those of braided-stream origin. 'Cross-bedding' on a very large (seismic) scale is referred to as clinoforms (see Sections 5.3.3.14 and 5.7).

Cross-stratification deserves careful observation in the field as it is a most useful structure for sedimentological interpretations, including palaeocurrent analysis (see Section 7.3.1). Table 5.3 provides references to the appropriate sections for further information.

5.3.3.1 Cross-lamination and cross-bedding

Cross-stratification forms either a single *set* or several/many sets (then termed a *coset*) within one bed (Figure 5.13). On size alone, the two principal types of cross-stratification are *cross-lamination*, where the set height is less than 6 cm and the thickness of the cross-laminae is only a few millimetres; and *cross-bedding*, where the set height is generally greater than 6 cm and the individual cross-beds are many millimetres to a centimetre or more in thickness.

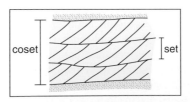

Figure 5.13 *A coset with three sets of cross-stratification.*

122

Table 5.3   *Cross-stratification: how to deal with it.*

1. Measure: (i) set thickness, (ii) coset thickness, (iii) cross-bed/cross-lamina thickness, (iv) maximum angle-of-dip of cross-strata, (v) direction of dip of cross-strata for palaeocurrent analysis (see *Chapter 7*)

2. Ascertain whether cross-lamination (set thickness <6 cm, cross-laminae <few millimetres thick) **or cross-bedding** (set thickness usually >6 cm, cross-beds more than few millimetres thick) (*Section 5.3.3.1*).

3. If cross-lamination: (i) Examine shape of foresets: tabular or trough (*Section 5.3.3.2, Figures 5.16, 5.17 and 5.38*). (ii) Is it climbing-ripple cross-lamination (*Section 5.3.3.4*), are stoss sides erosional surfaces or are stoss-side laminae preserved (*Figure 5.18*)? (iii) Is it current-ripple or wave-ripple cross-lamination? Look for form-discordant laminae, draping foreset laminae, undulating or chevron lamination – all of which characterise wave-formed cross-lamination (*Section 5.3.3.5, Figure 5.19*). (iv) Are there mud drapes giving flaser bedding or interbedded mud horizons giving wavy bedding or are there cross-laminated lenses in mudrock giving lenticular bedding (*Section 5.3.3.6, Figures 5.20–5.22*)?

4. If cross-bedding: (i) Examine shape of cross-bed sets: trough, tabular or wedge-shaped? *Section 5.3.3.2, Figure 5.16*. Are there any master bounding surfaces present? *Section 5.3.3.2; Figure 5.15*. (ii) Examine foresets: planar or trough beds, angular or tangential basal contact? *Figures 5.14 and 5.16*. (iii) Check bottom-sets of cross-beds for cross-laminae: co-flow or back-flow? *Section 5.3.3.2, Figure 5.14*. (iv) Examine texture of sediment: note distribution of grain-size, look for sorting and grading of sediment in cross-beds and alternations of coarse and fine beds. *Section 5.3.3.3, Figure 5.14*. (v) Look for internal erosion surfaces within cross-bed sets; are they reactivation surfaces? *Section 5.3.3.7, Figure 5.23*. (vi) Is there any evidence for tidal currents? Features indicating a tidal origin include: herring-bone cross-bedding; a bundling of the cross-beds (systematic variations in thickness of cross-beds and grain-size along the section); mud drapes on the cross-beds; lenses

(continued overleaf)

6. Fossils in the Field

7. Palaeocurrent Analysis

8. Facies Identification

Table 5.3 *(continued)*

of reversed flow cross-lamination (*Section 5.3.3.8, Figures 5.23–5.26*). (vii) Is there any evidence for storm waves? Is hummocky or swaley cross-stratification developed? Are cross-beds undulating with low-angle truncations? *Section 5.3.3.9, Figures 5.27–5.30*. (viii) If low-angle cross-beds, could it be antidune cross-bedding (*Section 5.3.3.15*) or beach lamination: very low-angle cross-beds in truncated sets? *Section 5.3.3.10, Figures 5.31 and 5.32*. (ix) If large-scale cross-bedding with high-angle dips, could it be aeolian? *Section 5.3.3.11, Figures 5.32 and 5.33*. Is pin-striped bedding present? *Figure 5.34*. (x) Look for low-angle surfaces within the cross-bedded unit; are they lateral accretion surfaces? *Section 5.3.3.12, Figures 5.3, 5.35 and 8.13*. (xi) Is the cross-bedding on a giant scale? Did it form through progradation of fan deltas or small 'Gilbert-type' deltas? *Section 5.3.3.13*. Are they clinoforms? *Section 5.3.3.14, Figures 5.36 and 5.83*.

5.3.3.2 Shape of cross-strata

Most cross-stratification arises from the down-current (or down-wind) migration of ripples, dunes and sand-waves when sediment is moved up the stoss-side and then avalanches down the lee-side of the structure. The shape of the cross-strata reflects the shape of the lee slope and this depends on the characteristics of the flow, water depth and sediment grain-size.

The steeply dipping parts of the cross-strata are referred to as *foresets* and they can have either angular or tangential contacts with the horizontal; in the latter case, the lower less steeply dipping parts are called *bottom sets* (Figure 5.14, also seen in Figures 5.6, 5.26 and 5.38). Cross-lamination, which may be upcurrent-directed (back-flow) or downcurrent-directed (co-flow), can be developed within the bottom sets of large-scale cross-beds as a result of ripples being formed in the dune trough (Figure 5.14).

The *upper bounding surface* of a cross-bed set is usually an erosion surface; this is the case with most sets of cross-lamination too. However, rarely, when much sediment is being deposited, stoss-side laminae can be preserved (see Section 5.3.3.4 and Figure 5.18).

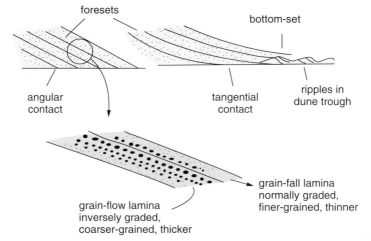

Figure 5.14 *Features of cross-stratification: basal contacts and internal texture.*

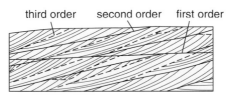

Figure 5.15 *Different orders of bounding surface in cross-bedded strata. Through-going bedding planes are referred to as master surfaces. No scale implied.*

In thicker sets of cross-beds (e.g. aeolian or shallow-marine sands), different hierarchies of bounding surface can be identified (Figure 5.15); some of these are *reactivation surfaces* (see Section 5.3.3.7 and Figure 5.23). Some horizontal bounding surfaces (first order, Figure 5.15) are laterally extensive bedding planes (*master surfaces*) and these may have an environmental significance (e.g. effects of rising watertable in aeolian sands or a major storm event on a shallow-marine shelf).

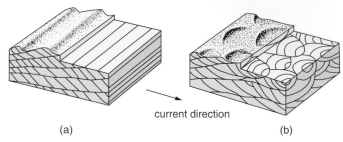

current direction

(a)   (b)

Figure 5.16 *Tabular (a) and trough (b) cross-stratification. In (a), cross-beds are generally planar with angular basal contacts, whereas in (b) cross-beds are scoop-shaped with tangential bases.*

Where the original bedform producing the cross-stratification is preserved (usually a ripple) and the cross-stratification is concordant with this shape, the cross-stratified unit is referred to as a *form set*.

The three-dimensional shape of cross-stratified units defines two common types: *tabular cross-strata*, where the inter-set boundaries are generally planar, and *trough cross-strata*, where the inter-set boundaries are scoop-shaped (Figure 5.16). Wedge-shaped cross-stratified units also occur. Tabular and wedge cross-bedding mostly consists of planar cross-beds that have an angular to tangential contact with the basal surface of the set. On the bedding surface, planar cross-beds are seen as straight lines. Trough cross-beds usually have tangential bases, and in bedding-plane view, the cross-beds have a nested, curved appearance.

Tabular cross-stratification is produced by straight-crested (i.e. two-dimensional) bedforms (Figure 5.16; also see Figure 5.38), whereas trough cross-stratification results from curve-crested (i.e. three-dimensional) bedforms (Figures 5.16 and 5.17).

Tabular cross-lamination is formed by straight-crested ripples; tabular cross-bedding is mainly produced by sand-waves, and also by straight-crested dunes. Trough cross-lamination is chiefly produced by linguoid ripples, and trough cross-bedding is mainly formed by lunate and sinuous dunes.

• Measure: the set and coset thickness of the cross-strata; the angle-of-dip of the cross-beds; and the direction of dip (or strike) of the cross-beds (for palaeocurrent analysis, see Chapter 7).

Figure 5.17 *Trough cross-beds. Notice that from this cross-section it is impossible to be sure of the palaeocurrent direction. Bioclastic shoreface grainstone. Pleistocene, Western Australia.*

5.3.3.3   Sorting in cross-beds and cross-bed types

Close observation of individual cross-beds will show variations in the grain-size distribution and may reveal different types of bed. When sand avalanches down the lee slope of the bedform, the deposited cross-beds show good sorting and inverse grading, that is, with coarser particles concentrated towards the outer/upper (downcurrent) part of each bed and towards the base (Figure 5.14). These are *grain-flow beds*. Currents (wind or water) transporting sand down the lee slope deposit *traction beds* that are normally graded. The coarser and thicker beds deposited by avalanching and traction often alternate with thinner beds of finer-grained material deposited from suspension. These are *grain-fall beds* (Figure 5.14). Where avalanching and traction were continuous (higher-energy conditions) then sorting within an individual cross-bed is less well developed and fine-grained beds are absent.

Finer sediment and plant debris are commonly concentrated in bottom sets of cross-beds, since these lightweight materials are carried over the ripple or dune crest and deposited in the trough. Thus the bottom sets of cross-beds may appear darker due to the presence of more clay and organic matter, as in Figure 5.26.

5.3.3.4   Climbing-ripple cross-lamination

When ripples are migrating and much sediment is being deposited, especially out of suspension, ripples will climb up the backs of those downcurrent to form climbing-ripple cross-lamination, also called

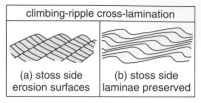

Figure 5.18 *Two types of climbing-ripple cross-lamination (ripple drift). In (a), sets of cross-laminae are bounded by erosion surfaces; in (b) stoss-side laminae are preserved so that cross-laminae are continuous.*

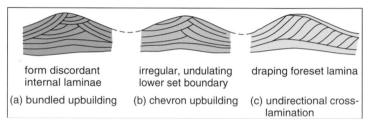

Figure 5.19 *Three types of internal structure of wave-formed ripples.*

ripple-drift. With rapid sedimentation, stoss-side laminae can be preserved so that laminae are continuous from one ripple to the next (Figure 5.18).

5.3.3.5   Wave-formed cross-lamination

The internal structure of wave-formed ripples is variable (Figure 5.19). Commonly, the laminae are not concordant with the ripple profile (i.e. the laminae are *form-discordant*). Two other features that distinguish wave-formed from current-ripple cross-lamination are irregular and undulating lower-set boundaries and draping foreset laminae (Figure 5.19).

5.3.3.6   Flaser, lenticular and wavy bedding

In some areas of ripple formation, the ripples of silt and sand move periodically and mud is deposited out of suspension at times of slack water.

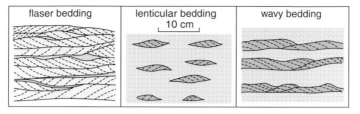

Figure 5.20 Sketches of flaser, lenticular and wavy bedding.

Figure 5.21 Flaser bedding, cross-laminated sandstone with thin drapes of mudrock. Section 20 cm high. Deltaic facies. Carboniferous, NE England.

Flaser bedding is where cross-laminated sand contains mud streaks, usually in the ripple troughs (Figures 5.20 and 5.21). *Lenticular bedding* is where mud dominates and the cross-laminated sand occurs in lenses (Figures 5.20 and 5.22). *Wavy bedding* is where thin ripple cross-laminated sandstones alternate with mudrock (see Figure 5.20). These bedding types are common in tidal-flat and delta-front sediments, where there are fluctuations in sediment supply or level of current (or wave) activity. Thin interbeds of sandstones and mudrocks are often referred to as *heterolithic facies*.

5.3.3.7 Reactivation surfaces
With some cross-bed sets, careful observation will show that there are erosion surfaces within them, cutting across the cross-strata

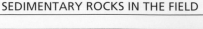

Figure 5.22 *Lenticular bedding: thin lenses of cross-laminated sand (sediment transport right to left) in dark-grey mudrock. Field of view 15 cm across. Outer shelf facies. Permian, Western Australia.*

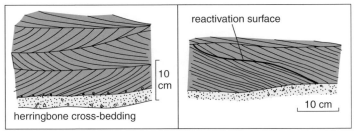

Figure 5.23 *Herring-bone cross-bedding and a reactivation surface in cross-bedding.*

(Figure 5.23). These *reactivation surfaces* represent short-term changes in the flow conditions that caused modification to the shape of the bedform. They can occur in tidal-sand deposits through tidal-current reversals or the effects of storms, in fluvial sediments through changes in river stage, and in aeolian sands through changes in wind strength or direction.

5.3.3.8 Tidal cross-bedding

There are several features of cross-bedding that indicate deposition by tidal currents. *Herring-bone cross-bedding* refers to bipolar cross-bedding, where cross-bed dips of adjacent sets are oriented in opposite directions (see Figures 5.23 and 6.6). Herring-bone cross-bedding is produced by reversals of the current, causing dunes and sand-waves to change their direction of migration. It is a characteristic but not ubiquitous feature of tidal sand deposits. Do check that the bipolar, herring-bone appearance is not due to a section through trough cross-bedding (Figure 5.17).

In many cases tidal cross-bedding is all unidirectional, since one tidal current is much stronger than the other. However, there may be subtle features to indicate a tidal origin: there may be *mud drapes* on cross-bed surfaces reflecting deposition from slack water during tidal current reversals (Figure 5.24; see also Figure 5.26); there may be thin lenses of ripples and cross-lamination within the cross-beds with a current direction opposite to that of the cross-beds (i.e. up the lee slope of the sand-wave/dune), indicating a weak, reverse-flow tidal current (Figure 5.24).

Some larger-scale tidal cross-strata show a regular pattern of varying cross-bed thickness and grain-size along the section, that is, through time (see, e.g., Figures 5.25 and 5.26); these *tidal bundles* reflect the

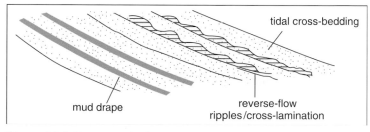

Figure 5.24 Features of tidal cross-bedding: mud drapes and reverse flow cross-lamination.

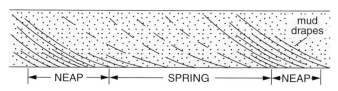

Figure 5.25 Tidal bundles in cross-beds: regular variations in the amount of mud/fine sand in the succession of cross-strata.

5. Sedimentary Structures

6. Fossils in the Field

7. Palaeocurrent Analysis

8. Facies Identification

Figure 5.26 *Cross-bedding (set 0.2 m thick) of tidal origin from a migrating sand-wave. Note the presence of mud drapes on the foresets and their regularity, reflecting the neap-spring lunar tidal cycle (see Figure 5.25). Shelf sandstone. Jurassic, Argentina.*

increasing and decreasing strength of tidal currents through the lunar cycle. You can measure up the thickness of the cross-beds and see how many days there were in the month! Reactivation surfaces (see Section 5.3.3.7 and Figure 5.23) and master bedding planes (see Section 5.3.3.2 and Figure 5.15) are also common in tidal-sand deposits from the effects of storms modifying and eroding the sand-wave/dune.

Tidal sandstones (or limestones) are usually well-sorted and rounded sands (texturally mature-supermature), although there may be thin pebbly lags/conglomeratic lenses; fossils/trace fossils may be present too. Compositionally, tidal sandstones are also typically mature-supermature.

5.3.3.9 Storm bedding: hummocky cross-stratification (HCS), swaley cross-stratification (SCS) and tempestites

Hummocky cross-stratification (*HCS*) and swaley cross-stratification (*SCS*) are two particular types of cross-stratification in sand-grade sediment, widely thought to be the result of storm waves and deposition in the transition zone between fairweather wave-base and storm wave-base and in the outer shoreface. *HCS* is characterised by a gently undulating low-angle (<10–15°) cross-lamination with the convex-upward part the hummock and the concave-downward part the swale (Figures 5.27–5.30). The spacing of the hummocks is several tens of centimetres to a metre or more, and in plan view they have a domal shape.

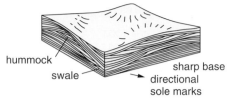

(a) hummocky cross-stratification
wavelength 0.5–5 m

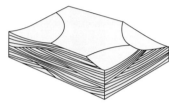

(b) swaley cross-stratification

Figure 5.27 *(a) Hummocky cross-stratification (HCS) and (b) swaley cross-stratification (SCS).*

Figure 5.28 *Hummocky cross-stratification in a mid-ramp bioclastic packstone. Wavelength 1 m. Upper Permian, NE England.*

Some hummocky beds show a succession of divisions: from basal (B), to planar-bedded (P), hummocky (H, the main part of the bed), flat-bedded (F), cross-laminated (X) to mudrock (M), reflecting a change from strong unidirectional currents (B, P) to oscillatory flow (storm waves: H, F, X) to deposition from suspension (M).

SEDIMENTARY ROCKS IN THE FIELD

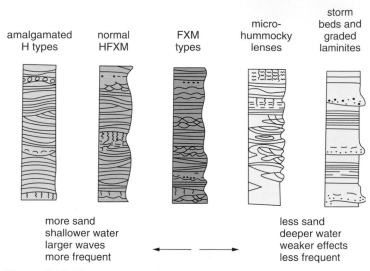

amalgamated H types | normal HFXM | FXM types | micro-hummocky lenses | storm beds and graded laminites

more sand
shallower water
larger waves
more frequent

⟵ ⟶

less sand
deeper water
weaker effects
less frequent

Figure 5.29 *The spectrum of storm deposits from beds dominated by hummocky cross-stratification (HCS) in a more proximal (shallower-water) area to storm beds ('tempestites') of a more distal (deeper-water) area. H = HCS, F = flat-bedding, X = cross-lamination, M = mud.*

Related to HCS is *SCS*, where hummocks are rare and the bedding mostly consists of broad concave-up laminae (Figure 5.27). Parallel lamination/flat-bedding with parting lineation (see Section 5.3.1.2 and Figure 5.7) is commonly associated. SCS is thought to form in higher-energy areas than HCS, in the inner transition to mid-shoreface zone.

Tempestites. Sandstone/limestone beds with HCS and SCS are one end of a spectrum of deposits resulting from storm deposition (Figure 5.29). Storm-wave deposits give way to storm-current deposits below storm wave-base, and these typically consist of sharp-based (with sole structures), graded and cross-laminated beds (0.01–0.5 m thick), interbedded with outer shelf mudrocks. Storm current beds are often referred to as '*tempestites*'. Some such beds are concentrations of fossils (see Section 6.2.2).

Figure 5.30 *Storm beds with undulating lamination (hummocky cross-stratification, HCS). Notice the parallel-sided nature of the beds in the lower part and the increased channelling higher up, indicating an increase in storm activity. This could have been the result of shallowing through progradation of the shoreline. Meso-Proterozoic limestones, eastern Ghats, India.*

5.3.3.10 Beach cross-stratification

Siliciclastic and carbonate sands deposited on a beach (foreshore) of moderate to high wave activity are characterised by a very-low-angle planar cross-stratification, arranged into truncated sets (Figures 5.31 and 5.32). The low-angle flat-bedding is typically directed offshore, but shoreward-directed bedding may also be present through sand deposition on the landward side of a beach berm. Boundaries between sets represent seasonal changes in the beach profile. The beds, which may show reverse grading, are formed by wave swash-backwash and commonly possess primary current lineation (see Section 5.3.1.3 and Figure 5.7). In addition, shallow channel structures may be present, oriented normal to the shoreline and cutting the flat-bedding at a low angle, which were produced by rip currents. Lenses of cross-laminated and/or cross-bedded sand may occur within the flat-bedded beach sand from the development of ripples (especially wave-formed), ridges and berms in the foreshore environment.

The texture and composition of the sediment may help to confirm a foreshore origin (see Sections 4.2 and 3.2.1); beach sandstones are typically quartzitic arenites with well-sorted and rounded grains. Burrows

Figure 5.31 *Beach-foreshore bedding consisting of truncated sets of low-angle, parallel-laminated sand (with parting lineation). Notice the horizontal and parallel nature of the lamination in the original shoreline, strike direction on the left face, compared with the dip-section ahead. Field of view 2 m across. Bioclastic grainstone. Pleistocene, Mallorca.*

Figure 5.32 *Foreshore facies in lower part of cliff dipping at a low angle seawards (to the left), overlain by aeolian facies, the bottom sets of large-scale cross-bedding dipping onshore (to the right). The vertical pipes are rhizocretions, calcretised tree roots (see Section 5.5.6.2). Bioclastic grainstone. Pleistocene aeolianite, Western Australia.*

and fossils may be present, and also heavy mineral layers, including placer beds of magnetite or ilmenite grains (see Section 3.7). There may be interbedded lenses and beds of conglomerate, with well-sorted and rounded pebbles, which may show imbrication (see Section 4.4) and might be bored.

Limestones deposited in a beach environment may contain *keystone vugs* (see Section 5.4.1.2), millimetre-sized holes, mostly now filled with calcite, formed through the entrapment of air in the sand through the rising and falling tide.

5.3.3.11 Aeolian cross-bedding

Compared with cross-bedding of subaqueous origin, cross-bedding produced by wind action generally forms sets that are much thicker and the cross-beds themselves dip at higher angles (Figure 5.33). Sets of aeolian cross-beds are typically several metres (up to 30 m) in height. Cross-beds can be trough or planar in shape and they most commonly have tangential bases. Foresets commonly dip up to angles in excess of 30°. By contrast, cross-bedding formed subaqueously is generally less than 2 m thick and cross-bed dips are generally less than 25°.

Figure 5.33 *Aeolian cross-bedding: two sets (the lower, 2 m thick; the upper, 4 m thick, between the arrows) of large-scale cross-bedded, yellow, clean, quartz arenite with high angles of dip. The desert sandstone rests unconformably on Carboniferous coal measures, and is overlain by an organic-rich dolomitic mudstone and then dolomite from a marine incursion into the basin. Height of exposure 12 m. Permian, NE England.*

Figure 5.34 *Aeolian facies: pin-stripe lamination from wind-ripple migration. Bioclastic grainstone. Field of view 30 cm across. Pleistocene aeolianite, Western Australia.*

Within aeolian cross-bed sets, there are commonly major horizontal erosion surfaces cutting through the cross-beds (master bedding surfaces, see Figure 5.15), which represent periods of deflation and changes in the position of the watertable. The cross-beds themselves commonly show reverse grading from avalanching (grain-flow deposits), and normal grading from traction flow; these coarser layers may alternate with finer, thinner beds deposited from suspension (grain-fall deposits) (see Figure 5.14). A further type of aeolian bedding is a *pin-stripe lamination*, millimetre-thick, which is formed through the migration of wind-ripples (Figure 5.34). Aeolian sandstone successions typically consist solely of large-scale cross-bedded sets, although there may be associated flat-bedded sandstones from interdune areas, and interbedded thin conglomerates and cross-bedded sandstones of fluvial origin.

If an aeolian origin is suspected, then also look at the composition and texture of the sediment (see Sections 3.2.1 and 4.2). Aeolian sandstones typically consist of well-sorted and well-rounded medium-sand quartz arenites. The quartz grains may show frosting (a dull lustre) and mica is generally absent. They may also be red. Aeolian limestones also occur (referred to as *aeolianites*, see Figures 5.32 and 5.34), composed of sand-sized bioclastic material, forming along coasts of high carbonate productivity.

5.3.3.12 Lateral accretion surfaces (epsilon cross-bedding)
Within cross-bedded channel sandstones there can sometimes be discerned a larger-scale low-angle bedding, oriented normal to

Figure 5.35 *Lateral accretion surfaces giving a large-scale (epsilon) cross-stratification, directed to the left, within a small channel, which in fact is cut by a larger channel to the left (channel bases indicated by dashed white lines). Palaeocurrent direction towards observer, as deduced from smaller scale cross-bedding within the sandstone units and the orientation of the channel-fill itself. Above the two channels is a coal seam (dashed green line), and then a coarsening-upward mudrock to sandstone unit from a small delta-lobe progradation. This is itself overlain by another coal seam (dashed green line), with a yellowish colour in places from the presence of jarosite, an iron sulphate weathering product of pyrite. Triangles indicate fining upwards and inverted triangle indicates coarsening upwards. Cliff is 8 m high. Proximal deltaic facies. Mid-Carboniferous, NE England.*

medium/smaller-scale cross-stratification (Figures 5.3 and 5.35). The surface cuts across the unit at a low angle (5–10°) and becomes asymptotic towards the base of the channel sandstone. This *epsilon cross-bedding* forms through lateral migration of the channel and represents the successive growth and lateral accretion of point bars. These surfaces are generally a metre or more in height and continue laterally for several to more than 10 metres. *Lateral accretion surfaces* are typical of meandering river channel sandstones but they can also occur in delta distributary-channel and tidal-channel deposits.

5.3.3.13   Small-delta/fan-delta cross-bedding

Where small, simple deltas build into lakes and lagoons (these are often referred to as 'Gilbert- type' deltas), very large-scale cross-bedding can develop, which represents the prograding front of the delta (delta slope).

The thickness of the cross-bedded unit (a few metres, exceptionally up to tens, but even hundreds of metres) reflects the depth of water into which the delta was advancing. For a very high cross-bed set, the term *clinoform* can be used (see Section 5.3.3.14).

Foresets dip at angles up to 25° and usually consist of sand, passing down into much finer-grained and well-developed bottom-sets of silt and clay, deposited largely from suspension in front of the delta. Top-sets are also well developed and can consist of lenticular gravels, sands and finer sediments, deposited by streams on the delta top, where they may be reworked by waves.

Small-delta cross-bedding is identified by the presence of well-developed top-sets and bottom-sets (the former contrasting with cross-bedding formed from dunes and sand-waves) and the presence of one thick set. These small deltas occur as a wedge or fan in marginal-lacustrine or marginal-marine settings.

5.3.3.14 Very large-scale cross-strata and clinoforms

In some situations, high cliffs along coasts or in mountainous areas, very large-scale cross-strata can be observed (Figures 5.36, 5.84 and 5.86). If these are on a seismic scale (>50 m thick), then the term *clinoforms* can be used. These features are commonly developed at carbonate platform margins and adjacent to reefs, where the dipping beds will be made of shallow-water material, fragments of reef-rock,

Figure 5.36 *Clinoforms: large-scale, gently dipping, prograding beds of limestone (slope facies) that are offlapping from a carbonate platform and downlapping onto deeper water mudrocks. Cliff 50 m high. Bioclastic wackestones. Cretaceous, Pyrenees, Spain.*

bioclasts and ooids, for example. The angle of dip with these debris beds varies from a few degrees to around 30°, depending largely on the grain-size of the sediment available (lower angle for finer sediment). There may be quite large blocks within the dipping beds or at their base.

If accessible, look at the geometry of the clinoform beds (planar, sigmoidal, oblique), the thickness, sediment grain-size and the internal structure of the beds. Are there any patterns (cycles) in the bed thickness, such as thickening, thinning or bundling? See Section 8.4.3.

5.3.3.15   Antidune cross-bedding

This is a rare but important structure since it indicates high flow velocities in the upper flow regime. Antidunes are low-amplitude bedforms in sand-grade sediment that migrate upstream through deposition of sediment on the upstream-facing slope of the bedform. Antidunes can be observed on modern beaches, in the backwash zone and in rivers, recognised by standing waves and breaking waves moving upstream against the flow. The cross-bedding formed by the migration of antidunes is directed upcurrent, so other evidence of flow direction is required (e.g. flute marks on the base of the bed) to be certain that the supposed antidune cross-bedding is oriented against the flow. This type of cross-bedding generally has a lower angle of dip and is less well defined than that formed by lower flow velocity subaqueous dunes. Antidune cross-bedding is known from turbidite and fluvial sandstones (but it is very rare), and pyroclastic surge deposits.

5.3.3.16   Cross-bedding in conglomerates

This is usually present in single sets, with set heights in the range of 0.2–2 m. Cross-beds, generally planar and low-angle, occur in tabular, wedge- and lenticular-shaped units. Cross-bedding is common in conglomerates deposited in fluvial environments (braided streams and stream floods) where it forms from the downstream migration of bars at high stage.

5.3.4   Graded beds

These beds show grain-size changes from bottom to top. The most common is *normal graded bedding*, where the coarsest particles at the base give way to finer particles higher up (Figure 5.37). The upward decrease in grain-size can be shown by all particles in the bed or by the coarsest particles only, with little change in the grain-size of the matrix.

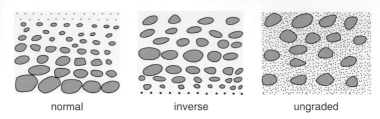

normal   inverse   ungraded

Figure 5.37 *Different types of graded bedding.*

Composite or multiple-graded bedding is where there are several graded units within one bed.

Less commonly, *reverse* (or *inverse*) *grading* is developed, where the grain-size increases upwards (see Figure 5.37). This can occur throughout a bed, or more commonly it occurs in the bottom few centimetres of the bed, with normal grading following. Reverse grading may only affect the coarse particles. Graded bedding can be observed (and measured) in conglomerates with no difficulty (see Section 4.6) and in sandstones with the aid of a hand-lens.

Normal graded bedding usually results through deposition from waning flows; as a flow decelerates so the coarsest (heaviest) particles are deposited first and then the finer particles. Such graded bedding is typical of turbidity-current and storm-current deposits (see, e.g., Figures 8.22 and 8.23). Composite graded bedding is usually the result of pulses in the current.

Reverse grading can arise from an increasing strength of flow during sedimentation, but more commonly from grain dispersion and buoyancy effects. It commonly occurs in the deposits of high-concentration sediment-water mixtures. Laminae deposited on beaches by swash-backwash are commonly reverse graded, as are cross-beds deposited by avalanching and grain flow (see Figure 5.14). Reverse grading can occur in the lowest parts of sediment gravity-flow deposits such as grain-flow and debris-flow deposits.

5.3.5 Massive beds

Massive beds have no apparent internal structure. It is first necessary to ascertain that this really is the case and that it is not simply due to surface weathering or a uniform grain-size. Blocks collected in the

142

field and then cut and polished or etched in the laboratory may show that structures are indeed present (lamination, bioturbation, etc.). Other techniques that may bring structures to light include:

- staining the surface, as with methylene blue, which highlights the organic matter; or Alizarin red S + potassium ferricyanide, which picks out dolomite/calcite and iron-rich/iron-poor carbonate;
- covering the surface in a thin oil (very useful for chalks);
- or using X-radiography at the local hospital (cut 0.5 cm-thick slabs of the rock and ask nicely!).

If the bed really is structureless then this is of interest and attempts should be made to deduce why. The two alternatives are that it was deposited without any structure or that the depositional structure was subsequently obliterated by such processes as bioturbation (see Section 5.6.1), recrystallisation, dolomitisation and dewatering (see Section 5.5). Where an original structure has been destroyed, careful examination of the bed, or again of cut and stained blocks, may reveal evidence for this; wisps of lamination may remain, the sediment may appear churned and homogenised.

Truly massive beds deposited in this condition mostly arise through rapid sedimentation ('dumping'), where there was insufficient time for bedforms to develop. Massive bedding is a feature of some turbidity-current and grain-flow sandstones, and debris-flow deposits, and occurs in some fluvial sandstones; see Figures 5.38 and 4.10, for examples.

5.3.6 Shrinkage cracks (mudcracks) and polygonal structures

These are present in many fine-grained sediments, especially mudrocks and lime mudstones, and most form through desiccation on exposure, causing shrinkage of the bed or lamina(e) and thus cracking. Many *desiccation cracks* define a polygonal pattern on the bedding surface (Figure 5.39), although they may also be seen on bed undersurfaces. Polygons vary enormously in size, from millimetres to metres across. Several orders of crack pattern may be present. Sediment clasts can be liberated by desiccation and lead to intraformational and edgewise conglomerates.

Sediments also crack subaqueously. S*yneresis cracks* form through sediment dewatering, often resulting from salinity changes or osmotic

Figure 5.38 *Massive bedding within a channel cutting into fluvial planar cross-bedded sandstones with overturning at the tops of the beds. Depth of channel 2 m. Braided stream fluvial facies. Lower Carboniferous, NE England.*

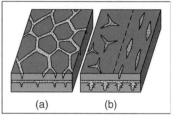

Figure 5.39 *Shrinkage cracks. (a) Formed by desiccation, typically complete polygons, can be straight-sided as shown, or less regular. (b) Formed through syneresis, typically incomplete with either bird's foot or spindle shape. In (a), the cracks are depicted as having suffered little subsequent compaction and so appear V-shaped in section; in (b) the crack fills are ptygmatically folded through compaction.*

effects. Syneresis cracks are characterised by an incomplete polygonal pattern. The cracks are often trilete or spindle-shaped (Figures 5.39 and 5.40) and can be mistaken for trace fossils or evaporite pseudomorphs (and vice versa).

Desiccation and syneresis cracks are usually filled with coarser sediment, seen in vertical section as wedges, although these can be later deformed and folded through compaction. Desiccation cracks indicate

Figure 5.40 *Syneresis cracks in muddy limestone. Shelf facies. Late Precambrian, Montana.*

subaerial exposure and so are common in sediments of marine and lacustrine shorelines, and river floodplains. Syneresis cracks mostly occur in shallow sublittoral lacustrine deposits.

Possibly related to syneresis cracks are *molar-tooth structures* (they look like their name!) – compacted cracks filled with fine-grained calcite occurring in lime mudstones and dolomites. They are common in Precambrian strata although their origin is controversial. They may relate to seismic activity, or not.

Polygonal cracks may also develop in carbonate sediments through early cementation and expansion of the surface crust. Tepee structures are commonly associated (see Section 5.4.2).

5.3.7 Rainspots

Rainspots are small depressions with rims, formed through the impact of rain on the soft exposed surface of fine-grained sediments. In some cases the spots are asymmetrical and you can tell which way the wind was blowing. They mostly occur in desert playa and lacustrine shoreline sediments.

5.4 Depositional Structures of Limestones (Including Dolomites)

This section describes sedimentary structures that are more commonly found in limestones rather than siliciclastic sediments.

5.4.1 Cavity structures

Many limestones contain structures that were originally cavities (or may still be so) but have since been filled with sediment and/or carbonate cement, in many cases soon after deposition. These include geopetal structures, fenestrae (including birdseyes), stromatactis, sheet cracks, neptunian dykes, caves, potholes and vugs.

5.4.1.1 Geopetal structures

This term can be applied to any cavity, mostly but not only in limestones, filled with internal sediment and cement (normally sparry calcite). The cavity-fill is a useful way-up indicator (the white sparite at the top), and the surface of the internal sediment provides a 'spirit-level', showing the position of the horizontal at the time of deposition. Geopetal structures commonly form beneath shells ('umbrella' structures), within skeletal grains and in synsedimentary cavities (Figures 5.41 and 5.42).

Careful measurement of geopetal structures can show that a series of limestones had an original depositional dip (Figure 5.42). This is usually the case with fore-reef limestones and the flank beds to mud mounds and patch reefs (as in Figure 5.83).

Figure 5.41 *Geopetal structure: a cavity filled with several layers of internal sediment (pink, at base of cavity), showing the way-up, and fibrous calcite (white), and calcite spar (brown) in cavity centre. Cavity 2 cm high. Microbial boundstone. Devonian, Windjana Gorge, Western Australia.*

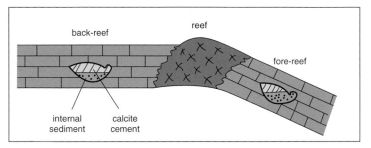

Figure 5.42 *Sketch of geopetal structures in back-reef facies, where sediment was deposited on a horizontal surface, and in fore-reef facies, where sediment was deposited on a slope.*

5.4.1.2 Fenestrae (including birdseyes)

These are cavity structures, usually filled with sparite, occurring in micritic, commonly peloidal limestones and dolomites. There are three common types of fenestra: equant to irregular fenestrae (birdseyes), laminoid fenestrae and tubular fenestrae.

Birdseyes are typically a few millimetres across, equant in shape and formed through gas entrapment and desiccation in tidal-flat carbonate sediments (Figure 5.43).

Figure 5.43 *Birdseyes (fenestrae) filled with sparite (grey calcite crystals) within peloidal limestone. Sutured stylolites also present. Field of view 4 × 2 cm. Tidal-flat facies. Devonian, Geikie Gorge, Western Australia.*

Laminoid fenestrae are elongate cavity structures, thin sheet cracks in effect, parallel to the stratification. They usually occur between microbial laminae, and result from desiccation and parting of the layers in a microbial mat. They are typically a few millimetres in height, and several centimetres in length.

Tubular fenestrae are generally oriented normal to the bedding and are a few millimetres across and several centimetres long. They may branch downwards. These structures are mostly burrows or rootlets. Some may be filled with sediment rather than cement (sparite).

Keystone vugs are similar to irregular fenestrae (birdseyes) in shape (equant) and size (few millimetres across) but occur in more grainy rocks (bioclastic and oolitic grainstones). These are typical of beach sediments and form by the trapping of air in the sand.

5.4.1.3 Stromatactis

This is a specific type of cavity that is characterised by a smooth floor of internal sediment, an irregular roof and a cement filling, usually isopachous (equal thickness) fibrous calcite (grey) followed by drusy sparry calcite (white) (Figure 5.44). Stromatactis is common in mud-mound limestones (massive lime mudstones/biomicrites, mostly

Figure 5.44 *Stromatactis cavities (flat floor of internal sediment, irregular roof, fibrous and drusy calcite fill) within lime mudstone. Field of view 15 cm across. Microbial boundstone (mud mound). Devonian, Belgium.*

of Palaeozoic age), but its origin is unclear. Possible explanations include: sediment dewatering, local seafloor cementation and scouring of sediment, and dissolution of sponges.

5.4.1.4 Sheet cracks and neptunian dykes

These cavities are either parallel to or cutting the bedding. Both can vary considerably in size, particularly the neptunian dykes, which can penetrate down many metres. They have usually formed by cracking of the lithified or partially lithified sediment and opening up of cavities. They can be filled either with sediment of similar age and lithology to the host sediment, or with quite different and much younger material. Sheet cracks and neptunian dykes can form in several ways, but penecontemporaneous tectonic movements, early compaction and settling, or slight lateral/downslope movement can all cause cracking and fissuring of the limestone mass.

5.4.1.5 Karstic cavities and karstic breccias

Structures somewhat similar to neptunian dykes, but often on a larger scale, can develop within a limestone mass when brought into contact with meteoric water through uplift or a major sea-level fall. Dissolution (karstification) of the limestone can result in cave systems being formed, which will vary from narrow vertical/subvertical channels (*potholes*) through to large *caverns*. These cavities usually cut across the bedding, although they may locally follow the stratification, and the walls are smooth to undulating, rather than planar. *Flowstone* (layers of fibrous calcite) may coat the walls of these karstic cavities, and *speleothems* (stalactites/stalagmites) may also be present. Extensive cave systems develop close to the groundwater table (which may be perched) and in the capillary zone above.

In most cases palaeokarstic cavities are subsequently filled with sediment, and this can range from non-marine red and green marls, sandstones and conglomerates (which may contain vertebrate fossils), deposited by subterranean streams and percolating in from the land surface above, through to marine sediments washed in during a subsequent transgression.

A variety of distinctive *breccias* are associated with palaeokarsts and occur upon palaeokarstic surfaces (see Section 5.4.2). They form from the fracturing of limestone and collapse of caves and so consist of angular clasts of the host limestone with textures varying from fitted fabrics showing little displacement of clasts (*crackle breccia*), to a more clast-supported breccia where some movement has taken place (collapse breccias), through to reworked clast-supported breccio-conglomerates where material has been moved along in subterranean streams.

5.4.2 Palaeokarstic surfaces

Palaeokarstic surfaces are formed through subaerial exposure and meteoric dissolution (karstification) of a limestone surface, generally under a humid climate. They usually have an irregular topography (Figure 5.45), possibly with potholes and cracks descending down several metres (see Section 5.4.1.5 above), which developed beneath a soil cover. Upon the surface there may still be a soil (now a palaeosoil – red/grey mudrock ± rootlets) or this may have been removed during a subsequent marine transgression so that marine facies rest directly on the exposure surface. Volcanic ash may occur on the surface (see Section 3.11). Breccias of the host rock may also occur upon the karstic surface.

The top surfaces of limestone beds should be examined carefully for evidence of karstification. Metre-scale shallowing-upward limestone cycles are commonly capped by palaeokarstic surfaces, indicating subaerial exposure. However, remember that limestone bedding planes commonly show the effects of pressure dissolution, which can produce

Figure 5.45 *Palaeokarstic surface, now the bed of the stream. Note irregular, potholed surface, overlain by flat-bedded limestone. Shelf bioclastic wackestones. Lower Carboniferous, Yorkshire Dales, England.*

sharp, undulating through to irregular, pitted and sutured surfaces, not unlike palaeokarsts (see Section 5.5.7).

Associated with some palaeokarstic surfaces are *laminated crusts*. These would normally occur on the top of the limestone bed, although they may be cut by the dissolution surface. They typically consist of pale-brown, grey or reddened micritic limestone with a poorly defined lamination; small tubes may be present, which were the sites of rootlets. These crusts are pedogenic in origin and they may be calcified root-mats or inorganic/microbial precipitates. In addition, there may be other indicators of soil processes: calcrete nodules, black pebbles, pisoids/vadoids and rhizocretions (see Section 5.5.6.2).

Palaeokarstic surfaces are important since they indicate subaerial exposure. The degree of their development does depend on the climate and the duration of the emergence.

5.4.3 Hardgrounds

These structures are present in limestones where there has been synsedimentary cementation so that the sediment was partly or wholly lithified on the seafloor. The top surface of a hardground usually provides the best evidence for a cemented seafloor. A hardground surface is usually encrusted by sessile organisms such as oysters, serpulid worms and crinoids, and penetrated by borings of such organisms as annelids, lithophagid bivalves and sponges (Figures 5.46 and 5.47, also see Section 5.6.2). Beneath hardground surfaces there may be horizontal cavities, formed from currents scouring beneath the cemented seafloor, where fossils may be present, as well as cements.

Many hardground surfaces themselves are planar, having been subjected to corrasion (erosion through sand being moved across the surface), so that borings and fossils in the sediment may be truncated. Other hardgrounds, generally in deeper-water limestones and chalks, have a more irregular surface with some relief where a degree of seafloor dissolution has taken place. Hardgrounds commonly have a nodular appearance, and this may be related to burrows and bioturbation that formed before the sediment was lithified. Hardgrounds may be impregnated with iron minerals and/or phosphate and there may be iron oxide or phosphatic crusts on the hardground surface. Grains of glauconite may be present too. Hardgrounds can commonly be traced for many tens (even hundreds) of metres.

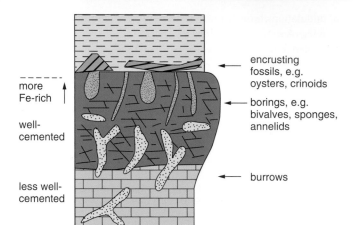

Figure 5.46 *Features of a hardground, generally indicating a pause in sedimentation and seafloor cementation.*

Figure 5.47 *View of hardground surface showing encrusting oysters and borings (circular holes). Thumb for scale. Oolitic grainstone. Jurassic, W England.*

152

Beds partially lithified on the seafloor are referred to as *firmgrounds* (see Section 5.6.2 and Table 5.11).

Hardgrounds are not that common but they do give important environmental and diagenetic information. They generally form during times of reduced or negligible sedimentation and they may coincide with periods of relative sea-level rise.

5.4.4 Tepee structures

As a result of synsedimentary cementation of carbonate sediments, the cemented surface layer can expand and crack into a polygonal pattern. Intraclasts can also be generated. The cemented crust can be pushed and buckled up to form *tepees* (pseudo-anticlines), and where cracked the crust can even be thrust over itself. Cavities can develop beneath the cemented surface layer and be filled with sediment and further cement. Tepee structures can develop in shallow subtidal sediments, in conjunction with hardgrounds, but more commonly they are found in tidal-flat carbonates. In the latter case, microbial laminites, desiccation cracks, pisoids (vadoids) and dripstone cements are usually associated. Thus, most tepees indicate subaerial exposure and marine diagenesis, and with prolonged exposure complex mega-tepee horizons can form.

5.4.5 Microbialites: microbial laminites, stromatolites and thrombolites; also tufa

These are biogenic structures that have a great variety of growth forms. They develop through the trapping and binding of carbonate particles by a surficial *microbial mat* (formerly 'algal mat') mainly composed of cyanobacteria (previously called blue-green algae) and other microbes, and the biochemical precipitation of carbonate. Microbialites vary from laminated forms, generally called stromatolites, to non-laminated or crudely laminated forms termed thrombolites. They are very common in Precambrian carbonates but also occur in many Phanerozoic limestones, particularly those of peritidal origin.

Stromatolites vary from planar-laminated forms, also called *microbial laminites*, to domes (like cabbages) and columns (Figure 5.48). Laminae are generally a millimetre to several millimetres in thickness, and consist of micrite, peloids and fine skeletal debris. Cement layers may also occur there.

With planar stromatolites (i.e. microbial laminites, Figure 5.49), the microbial origin of the laminae is shown by small corrugations and

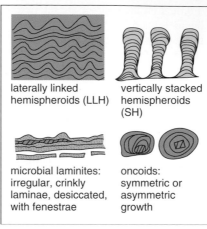

Figure 5.48 *Four common microbial (stromatolitic) structures: domal, columnar and planar stromatolites and oncolites.*

Figure 5.49 *Microbial laminites, with small domal structure. Field of view 30 cm across. Tidal-flat dolomites. Upper Permian, NE England.*

undulations and preferential thickening over small surface irregularities. The laminae are commonly broken and disrupted through desiccation, and they may show regrowth and encapsulations upon fragments of the microbial mat. Tepee structures may be present, and intraclasts are commonly generated from desiccation and exposure. Elongate cavities (laminoid fenestrae; see Section 5.4.1.2) are common in microbial laminites, and there may be thin grainstone beds (possibly of storm origin) intercalated. Microbial laminites are typical of tidal-flat limestones and dolomites and so may be associated with fenestral lime

Figure 5.50 *Microbial bioherm, 3 m across. Shelf dolomites. Upper Permian, NE England.*

mudstones and evaporites or their pseudomorphs (see Section 3.6). They may also form larger biohermal structures and buildups, up to many metres across (e.g. Figure 5.50).

Domal stromatolites generally have laminae continuous from one dome to another, and they can be several tens of centimetres in diameter or more. *Columnar stromatolites* are discrete structures, usually forming in higher energy locations, so that intraclasts and grains commonly occur between the columns. Figure 5.51 shows a cross-section

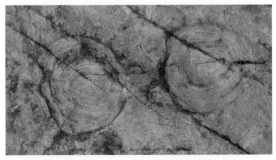

Figure 5.51 *Cross-section through columnar stromatolites, each 10 cm across. Shelf microbial dolomites. Meso-Proterozoic, Eastern Ghats, India.*

of columnar stromatolites. If these structures do not show a good lamination, but rather a clotted or peloidal, even pisolitic texture, then the term *thrombolite* would be more appropriate. The laminated versus non-laminated nature of microbialites is a reflection of the original microbial mat community (filamentous versus coccoid microbial forms) and environmental factors. Small cavities (fenestrae) are common within the domal-columnar microbialites.

A useful notation for describing stromatolites in the field is given in Figure 5.48. The morphology of stromatolites commonly changes upwards, as a response to environmental changes, and large stromatolite structures may consist of lower orders of domes and columns. This notation provides a convenient shorthand for describing such changes and variations. Stromatolites may form thin beds or they may constitute complex reefal-type structures. In Precambrian rocks, stromatolites are very diverse in shape and microstructure and many have been given form genus and species names, just like trace fossils.

Oncoids (also oncolites) are spherical to subspherical, unattached microbial structures, commonly with a crude concentric lamination (see Figure 3.8). The laminae may be asymmetric and discontinuous from growth on the top of the ball while it was stationary. These microbial structures can be mistaken for pisoids (also called vadoids) of pedogenic origin (see Section 5.5.6.2). Often looking similar to oncoids are *rhodoliths*, algal balls formed by calcareous algae, which are mostly red (coralline) algae.

Tufa is another type of microbialite, but forming in freshwater at ambient temperature near springs and groundwater seeps, in streams (creating barrages), and even on a lake floor if fresh waters emerge there. Tufa is a porous calcite deposit, often containing calcified plants and leaves, and forms from microbially induced and chemical calcite precipitation. Related to tufa is *travertine*, precipitated at warm- to hot-water springs, also calcite and related to microbial-fungal processes. *Sinter* is similar but it is made of siliceous material and usually occurs in association with volcanic activity.

5.4.6 Microbially induced sedimentary structures (MISS) in sandstones

Microbially induced sedimentary structures (MISS) occur in siliciclastic sediments and are produced by a cohesive microbial mat that once covered the sediment surface, causing its biostabilisation. The mat would have caused the baffling, trapping and binding of loose sand grains, but there would not have been any precipitation of carbonate, as with microbialites (see Section 5.4.5 above). MISS include wrinkle marks, and overturned and overfolded laminae, as a result of expansion of the microbial mat with included sandy layers, and centimetre-size domes and bumps from gas rising and being trapped beneath the mat. MISS occur especially in Precambrian sandstones because there were no grazing or burrowing organisms around at the time to disrupt the mat. See Recommended Reading for further information.

5.5 Post-Depositional Sedimentary Structures

A variety of structures are formed after deposition, some through mass movement of sediment (slumping and sliding) and others through internal reorganisation by dewatering and loading. Post-depositional physico-chemical and chemical processes produce stylolites, dissolution seams and nodules.

5.5.1 Slumps and slides and megabreccias

Once deposited, either upon or close to a slope, a mass of sediment can be transported downslope. Where there is little internal deformation of the sediment mass, more often the case with limestones, then the transported mass is referred to as a *slide*. Brecciation of the sediment mass can take place to produce large and small blocks.

Megabreccia is a term used for a deposit of large blocks. Some megabreccias are the result of fault activity during deposition and the erosion of fault scarps; others, notably those composed of limestone blocks and occurring in toe-of-slope locations, were commonly deposited during times of relative sea-level fall and collapse of a carbonate platform margin (see Figure 8.27). Examine the blocks in a megabreccia and see if there is any evidence for subaerial exposure of the sediments before their deposition into deeper water (e.g. look for dissolutional cavities filled with red soil, 'terra rossa'). See if the blocks have been bored or encrusted by sessile organisms (oysters, serpulids, etc.) or iron oxide or phosphorite (particularly on their upper surface), while in the deeper-water environment. Check to see if the blocks are really exotic and seemingly out of place.

Figure 5.52 *Slump-folded, deeper-water limestones. Disharmonic nature of folding shows that the deformation is synsedimentary. Height of cliff 10 m. Pelagic lime mudstones. Tertiary, Paxos, Greece.*

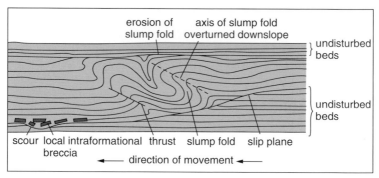

Figure 5.53 *Principal features of a slumped bed. Slumps can occur on a scale of centimetres to kilometres.*

Where a sediment mass is internally deformed during downslope movement, then the term *slump* is more appropriate. A slumped mass typically shows folding; recumbent folds, asymmetric anticlines and synclines, and thrust folds are common, on all scales (Figures 5.52 and 5.53). Fold axes are oriented parallel to the strike of the slope, and the direction of overturning of folds is downslope. It is thus worthwhile measuring the orientation of fold axes and axial planes of slump folds to ascertain the direction of slumping and so the palaeoslope. Slumps and slides range from metres to kilometres in size. Many are triggered by earthquake shocks.

The presence of a slump or slide in a succession can be deduced from the occurrence of undisturbed beds above and below, and a lower contact (the surface upon which the slump or slide took place), which cuts across the bedding (Figure 5.53). Be sure that lateral mass movement of sediment took place because somewhat similar convolutions and brecciations of strata can be produced by dewatering (see below) and other processes.

5.5.2 Deformed bedding

Deformed bedding, and terms such as disrupted, disturbed, convolute and contorted bedding, can be applied where the bedding, cross-bedding and cross-lamination produced during sedimentation have been subsequently deformed, but where there has been no large-scale lateral movement of the sediment itself (Figure 5.54).

Convolute bedding typically occurs in cross-laminated sediments, with the lamination deformed into rolls, small anticlines and sharp synclines. Such convolutions are commonly asymmetric and overturned in the palaeocurrent direction. Contorted, disturbed and disrupted bedding applies to less regular deformations within a bed, involving irregular

Figure 5.54 *Disturbed and contorted bedding. Field of view 2 × 1 m. Fluvial facies. Triassic, Broome, Western Australia.*

folding, contortions, disruptions and injections without any preferred orientation or arrangement. Wholesale or local brecciation of some beds can occur. *Overturned cross-bedding* affects the uppermost part of cross-beds and the overturning is invariably in the downcurrent direction (see Figure 5.38).

Deformed bedding can arise from a number of processes. Shearing by currents on a sediment surface and frictional drag exerted by moving sand are thought to cause some convolute bedding and overturned bedding. Dewatering processes such as fluidisation and liquefaction are often induced by earthquake shocks, in many cases associated with synsedimentary movements on faults.

5.5.3 Sandstone dykes and sand volcanoes

These are relatively rare structures but readily identifiable, the dykes from their cross-cutting relationship with bedding and fills of sand, and the sand volcanoes (formed where sand moving up a dyke reached the sediment surface) from their conical shape with a central depression, occurring on a bedding plane. Dewatering and the upward escape of water, usually initiated by earthquake shocks, is the cause of these structures.

5.5.4 Dish structures and dish-and-pillar structures

These consist of concave-up laminae (the *dishes*, Figure 5.55), generally a few centimetres across, which may be separated by structureless zones (the *pillars*). Dish structures and dish-and-pillar structures are formed by the lateral and upward passage of water through a sediment. Although not restricted to sandstones of a particular environment or depositional mechanism, they are common structures of sediment gravity flow deposits that occur in deepwater slope, fan and apron successions, and were often deposited quickly.

5.5.5 Load structures

Load structures are formed through differential sinking of one bed into another. *Load casts* are common on the soles of sandstone beds overlying mudrock, occurring as bulbous, rounded structures, generally without any preferred elongation or orientation (Figure 5.56). Mud can be injected up into the sand to form flame structures. Also as a result of loading, a bed, usually of sand, can sink into an underlying mud and break up into discrete masses, forming the so-called *ball-and-*

Figure 5.55 *Dish structures in sandstone. Field of view 10 cm across. Deltaic sandstone. Upper Carboniferous, NE England.*

Figure 5.56 *Load structures on underside of sandstone bed. Field of view 1 m. Deep-marne greywacke turbidite. Late Precambrian, Norway.*

pillow structure. On a smaller scale, individual ripples can sink into underlying mud, producing sunken ripples or sandstone balls.

Close attention should be paid to sandstone-mudrock junctions since it will often be found that gravity-loading has taken place (Figure 5.57).

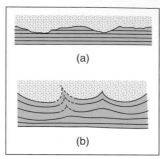

Figure 5.57 (a) Scoured surface at base of sand bed; truncation of mud laminae at contact. (b) Loaded surface, through sinking of sand into mud. Upward injection of mud (flame structure) and depression and contortion of mud laminae. A scoured surface can be deformed by loading.

5.5.6 Nodules

Nodules (also called concretions) commonly form in sediments after deposition; effectively most nodules are local patches of cementation. Two broad types are diagenetic nodules, formed during burial (shallow or deep), and pedogenic nodules, formed by soil processes.

5.5.6.1 Diagenetic nodules

Minerals commonly comprising nodules formed during diagenesis are fine-grained varieties of calcite, dolomite, siderite, pyrite, collophane (calcium phosphate), quartz (chert/flint; see, e.g., Figure 3.21) and gypsum-anhydrite. Calcite, gypsum-anhydrite, pyrite and siderite nodules of a few millimetres to a few tens of centimetres in diameter are common in mudrocks; chert nodules are more common in limestones, and calcite and dolomite nodules, in some cases of immense size (metres across), occur in sandstones. The latter are sometimes called *doggers* (Figure 5.58).

Nodules may be randomly dispersed or concentrated along particular horizons. Nodule shape can vary considerably from spherical to flattened, to elongate, to highly irregular. Some nodules nucleated around fossils and others formed within animal burrow systems; however, the majority are not related to any obvious pre-existing inhomogeneity in the sediment. Some nodules possess radial and concentric cracks, filled

Figure 5.58 *Fine-grained cross-laminated sandstone with large (2 m diameter) nodule ('dogger') where sandstone preferentially cemented by iron-rich dolomite. Also note the jointing: closer spacing within the dogger compared to the sandstone. Deltaic sandstone. Carboniferous, NE England.*

Figure 5.59 *Siderite nodule with septarian cracks filled by sediment (grey) and cement (white). Field of view 30 cm across. Marine mid-shelf facies. Upper Carboniferous, NE England.*

with crystals usually of calcite, or siderite or sediment (Figure 5.59). The cracks in these *septarian nodules* form through contraction (dewatering) of the nodules soon after their formation.

Geodes are a type of nodule with a hollow centre in which crystals have usually grown in towards the middle. Some geodes have formed through the dissolution of evaporite nodules (anhydrite especially), and they are usually composed of quartz, less commonly calcite or dolomite. These often have a 'cauliflower' appearance on the outside (see Figure 3.18).

A further particular type of nodule is one composed of calcite that exhibits a *cone-in-cone structure*. The calcite consists of fibrous crystals, several to 10 or more centimetres in length, which are arranged in a fanning, conical pattern oriented at right-angles to the bedding. Cone-in-cone structure generally develops in organic-rich mudrocks during burial, and the curious crystal form may relate to calcite crystal growth under high pore-fluid pressure during compaction.

Nodules can form at various times during diagenesis and burial; however, the majority of nodules in muddy sediments form during early diagenesis, before the main phase of compaction. Calcareous nodules are common in marine mudrocks generally, but they do form in soils as calcretes (see below), so are found in floodplain and lacustrine mudrocks too. Pyritic nodules are more common in organic-rich, muddy marine sediments (since seawater has an abundant supply of sulphate, which is reduced to sulphide by bacteria), whereas siderite is more common in organic-rich non-marine sediments.

As a result of the usual early diagenetic origin, lamination in host muddy sediment is deflected around the nodules (Figure 5.60), and the nodules can preserve the original lamina thickness. The amount of compaction that has taken place can be deduced from these

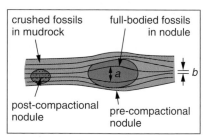

Figure 5.60 Pre-compactional (early diagenetic) and post-compactional (late diagenetic) nodules in mudrocks. An estimate of the amount of compaction can be obtained from early diagenetic nodules as $[(a-b)/a] \times 100\%$.

Table 5.4 *Nodules: what to look for.*

1. Determine composition of nodules and nature of host sediment.
2. Measure nodule size and spacing.
3. Describe shape and texture; are they forming within burrows?
4. Look for a nucleus (such as a fossil).
5. Try to determine when formed: (i) early diagenetic: contain full-bodied fossils, locally reworked, pre-compaction so host-sediment lamination is deflected. Or (ii) late diagenetic: contain crushed fossils, post-compaction so do not affect host-sediment lamination.
6. Check if calcrete (pedogenic nodules) *Section 5.5.6.2*.

nodules (Figure 5.60). Fossils and burrows in early diagenetic nodules are protected from compaction and so are unbroken and uncompressed, contrasting with compacted fossils in adjacent mudrocks.

Early diagenetic nodules forming close to the sediment-water interface can be exposed on the seafloor as a result of a storm, and then reworked, and perhaps encrusted and bored by organisms (the nodules act as local hardgrounds; see Section 5.4.3). Chert nodules in limestones (including flint nodules in chalk) and carbonate nodules in sandstones are also chiefly of early diagenetic origin. Late diagenetic nodules in muddy sediments formed after compaction are relatively rare; in these the lamination is undeflected from host sediment through the nodules (Figure 5.60).

A scheme for describing nodules is given in Table 5.4.

5.5.6.2 Calcretes (caliches): pedogenic nodules and limestones

Calcareous nodules and layers (calcrete or caliche) develop within soils of semi-arid environments where evaporation exceeds precipitation. In the geological record they are typically found in red-bed successions, in floodplain mudrocks especially, but they can also occur in marine clastic sediments, formed when the latter become emergent.

In the field, calcretes usually occur as pale-coloured nodules of fine-grained calcite, or rarely dolomite (*dolocrete*), a few to many centimetres in diameter, generally with a downward elongation (Figure 5.61),

Figure 5.61 *Calcrete (calcareous palaeosoil) consisting of elongate nodules of fine-grained calcite. Floodplain facies. Devonian, W England.*

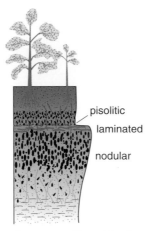

Figure 5.62 *Mature calcrete profile with densely packed elongate nodules – some will have formed around roots – and laminated calcrete.*

best developed in red mudrocks. They vary from randomly scattered to closely packed nodules, and may form a dense limestone bed as the calcrete develops and matures (Figure 5.62). Some calcrete has nucleated and grown around plant roots; *rhizocretion* and *rhizolith* is a general term for such precipitates and calcified roots and rootlets. These

Figure 5.63 *Rhizocretions: calcified plant roots. Field of view 3 m across. Floodplain mudrocks. Triassic, Devon, SW England.*

commonly taper downwards and usually branch (Figure 5.63; see also Figure 5.32).

The soil layer above a developing calcrete may be removed by erosion so that the calcrete then occurs at the land surface; a crust may develop there. *Duricrust* is a general term for such a lithified pedogenic surface layer. Apart from calcrete, duricrusts can be iron-rich (a ferricrete; see Section 3.7) or siliceous (silcrete; see Section 3.8).

Another form of pedogenic limestone is a *laminar calcrete* or *laminated crust* (see Section 5.4.3), where the laminae are usually slightly irregular in thickness and result from subtle variations in crystal size and colour. Some laminated calcretes are calcified root mats and so contain small tubules. These crusts develop upon nodular calcretes when the soil zone is fully cemented (plugged), and on limestone surfaces. They may also form around tree roots. In many calcretes there are peloids and pisoids (*vadoids*); these spherical grains are commonly irregularly laminated and are probably formed by microbial processes. They can be several centimetres in diameter.

Where calcretes have developed within coarser sediments, pebbles and grains may be split apart. In older more mature calcretes, fractures and veins of calcite may cross-cut a dense lithified calcrete and wholesale brecciation can occur, resulting in intraclasts and tepee structures (see Section 5.4.4). Cavities may be produced and filled with sediment, pisoids and cement.

Black pebbles are present in some calcretes, and can be reworked into basal lags. They are possibly the result of natural fires burning vegetation and blackening calcrete.

Calcretes are a useful palaeoclimatic indicator, but they also reflect lengthy periods of non-deposition (hundreds to thousands of years or more) during which the pedogenic processes operated. They also form at unconformities.

5.5.7 Pressure dissolution and compaction

As a result of overburden and tectonic pressure, dissolution takes place within sedimentary rock masses along certain planes to produce a variety of smooth to irregular surfaces (Figure 5.64). Pressure dissolution effects are commonly seen at the junctions of limestone beds and within limestones. This process probably begins quite early during burial, but is well developed after burial below many hundreds of metres. There are two major styles of pressure dissolution, leading to sutured and non-sutured seams. Pressure dissolution can result in a pseudobedding (Figure 5.65).

The sutured type are the well-known *stylolites*, which are generally bedding-parallel, although they can be at high angles to the bedding too. Insoluble material (chiefly clay) is concentrated along the seams (Figure 5.66). Stylolites occur as single sutured planes or as zones, or swarms; high- (several centimetres) and low-amplitude forms can be distinguished.

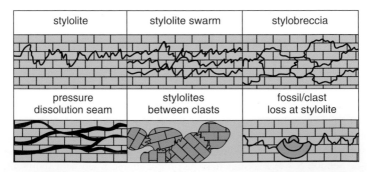

Figure 5.64 *Different products of pressure dissolution.*

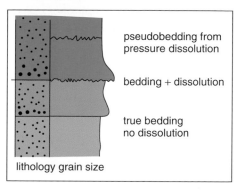

Figure 5.65 Bedding, pseudobedding and pressure dissolution planes.

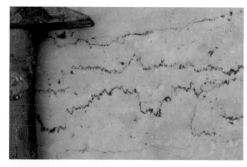

Figure 5.66 Stylolites in chalk. Field of view 20 cm across. Pelagic lime mudstone. Cretaceous, E England.

Non-sutured *dissolution seams* with little relief are also common, perhaps more so in argillaceous limestones. They are undulating, branching and anastomosing, and also occur singly or in seams (Figure 5.64). They can pass into cleavage planes in adjacent mudrocks. Nodular, lumpy and pinch-and-swell textures can be developed on the centimetre or decimetre scale.

Extreme pressure dissolution can give a limestone a brecciated appearance (*stylobreccia*) with a fitted fabric to the 'pebbles'. The term flaser limestone has been used for this rock-type. The occurrence of pressure dissolution can be demonstrated by the partial loss of fossils at or across stylolites. In some cases sutured contacts occur between fossils or pebbles in a limestone.

Stylolites also occur in sandstones, but they are not so common. They can occur between pebbles in conglomerates, leading to a pitting of the pebbles.

Compaction of muddy sediments begins soon after deposition, and the main effects are the crushing of fossils and reduction in thickness of deposited sediment by a factor of up to 10 (best seen where there are early diagenetic nodules in the mudrock; Section 5.5.6). Compaction coupled with pressure dissolution also enhances limestone-mudrock contacts.

5.5.8 Veins, beef and discontinuities (joints and fractures)

Although many *veins* are of purely tectonic origin and are mineralised through the passage of hydrothermal fluids, veins can form through compaction and burial diagenetic processes. The most common sedimentary veins are those composed of fibrous gypsum (satin spar), most of which occur in mudrocks and form through the hydration of anhydrite during uplift of evaporitic strata. Cross-cutting veins of calcite occur in some mature calcretes, karstified limestones, hardgrounds and tepee-ed and fractured supratidal carbonates, although in many cases these would be more appropriately called fractures and the mineral fills are cements.

Sheets of calcite parallel to bedding occur within some mudrocks, usually those rich in organic matter, and they generally consist of vertically oriented fibrous crystals. These calcite sheets are locally known as '*beef*' (!), and are related to cone-in-cone structure (see Section 5.5.6.1).

Joints and *fractures* in sedimentary rocks have mostly formed through tectonic stresses but in some cases they are related to depositional or burial processes: the cooling joints in some volcaniclastic deposits (see Section 3.11), joints in glacial diamictites from ice loading, the cleat in coal (see Section 3.10) and fractures in competent beds (such as well-cemented limestones and cherts) simply related to overburden pressure. There may be several sets of these discontinuities, and they may be of several generations too, cross-cutting and off-setting each other. The spacing of joints relates to

Table 5.5 *Features to note with fractures.*

Measure the orientation of the fractures, and their angle if they are not vertical; the fractures may indicate the regional stress field. Compare with fault and fold trends.

Measure the spacing of fractures; are they closely or widely spaced (*Table 5.6*)? See if there is a correlation of spacing with bed thickness.

How laterally persistent are the fractures? Are they arranged en echelon?

Do the fractures generate rock clasts? Are they blocky (equidimensional), or tabular (thickness much less than length or width) or columnar (height much greater than cross-section)?

Are the fractures filled by cement or sediment? Are they open or not? Are their walls smooth, rough, polished or slickensided? There may be alteration of the wall-rock adjacent to fractures from the passage of porefluids, as in local dolomitisation of a limestone, or in loss of colour in clastic red beds.

Table 5.6 *The spacing of discontinuities (fractures and joints) in sediments and sedimentary rocks.*

Term	Spacing
Very widely spaced	Greater than 2 m
Widely spaced	600 mm to 2 m
Medium spaced	200–600 mm
Closely spaced	60–200 mm
Very closely spaced	20–60 mm
Extremely closely spaced	Less than 20 mm

the thickness, competence and brittleness of the rock, with a closer spacing in more brittle rocks (see Figure 5.58 for an example). For the features to note with regard to joints and fractures see Table 5.5 and for the terms for spacing see Table 5.6.

Figure 5.67 *Dendrites – patterns looking like leaves but produced by manganese and iron oxide precipitation on a bedding plane. Field of view 10 cm across. Outer shelf/epeiric-sea lime mudstone. Jurassic, Germany.*

5.5.9 Dendrites

Dendrites are strange patterns, usually on bedding planes, resulting from the precipitation of manganese and iron oxides/hydroxides (Figure 5.67). Generally black and looking like fossil leaves, they are really something of a curiosity, with little sedimentological or diagenetic interest.

5.6 Biogenic Sedimentary Structures

Many structures are formed in sediments through the activities of animals and plants, in addition to the microbialites described in Section 5.4.5. Structures produced vary considerably from poorly defined and vague disruptions of lamination and bedding, to discrete and well-organised trace fossils (*ichnofossils*), which can be given a specific name. The texture produced by this biogenic activity is the sediment's *ichnofabric*. Trace fossils can often be interpreted in terms of the animal's activity that gave rise to the structure, but the nature of the animal itself can be difficult or impossible to deduce since different organisms may have had a similar mode of life. Also, one animal can produce different structures depending on its behaviour and on the sediment characteristics (grain-size, water content, etc.).

Burrow structures are commonly made by crustaceans, annelids, bivalves, echinoids and sea anemones, and surface trails and tracks by crustaceans, trilobites, annelids, gastropods and vertebrates. Structures somewhat similar to burrows can be produced by the roots of plants, although the latter may contain carbonised cores (see Section 5.6.3).

Care should be taken not to mistake modern biogenic markings, such as polychaete and sponge borings and limpet and periwinkle tracks, frequently seen on rocks in the intertidal zone, for trace fossils. Small holes and discolorations made by lichens on limestone can also look like ancient biogenic structures. Confusion can also arise with some sedimentary structures, such as syneresis cracks, tool marks, dewatering structures and diagenetic nodules, and spots and lineations in low-grade metamorphosed slates.

Some biogenic sedimentary structures on bedding planes are best seen in a directed low-angle light. Some major discoveries have been made at dawn (or dusk) when the sun is low in the sky. To save getting up early, examine the surfaces of loose slabs at different angles; it may be possible to see the structures better this way.

A scheme for the examination of trace fossils is given in Table 5.7.

5.6.1 Bioturbation

Bioturbation refers to the disruption of sediment by the activity of organisms and plants. The ichnofabric generated varies from scattered discrete burrow structures (commonly filled with sediment of a different colour, composition or grain-size) to completely disrupted sediment, which has a 'churned' appearance and a loss of depositional structure (e.g. Figure 5.68). Bioturbation may completely homogenise a sediment by mixing or it may lead to a nodular texture and segregations of coarser and finer sediment. A brecciated texture (*pseudobreccia*) can be produced by burrowing. In addition, burrows may be preferentially dolomitised or silicified. Mottled sediments, with colour variations, may also result from the bioturbation (*burrow mottling*). Some curious looking rocks have been produced by bioturbation and imaginative names given to them (e.g. leopard rock).

A *bioturbation index* can be devised to indicate the degree of sediment disruption or the percentage of ichnofabric in the sediment (see Table 5.8 and Figure 5.69); this index can be entered on a graphic log.

Table 5.7 *Trace fossils: how to describe them and what to look for.*

1. Sketch (or photograph) the structures; measure size, width and diameter, determine the bioturbation index. Note the facies context; check the sediment grain-size.

Points to note for trails and tracks, and burrows are:

2. *Trails and tracks* (on bedding surfaces or undersurfaces). (i) **Examine trail pattern**: note whether regular or irregular pattern, whether trail is straight, sinuous, curved, coiled, meandering or radial. (ii) **Examine trail itself**: if a continuous ridge or furrow note whether central division and any ornamentation (such as chevron pattern); with appendage marks or footprints, measure size and spacing (gait) of impressions, look for tail marks.

3. *Burrows* (occurring within beds, but also seen on bedding surfaces). (i) **Describe shape** and orientation to bedding; possibilities: horizontal, subvertical, vertical; simple straight tube, simple curved or irregularly disposed tube, U-tube. If branching burrow, note if regular or irregular branching pattern and any changes in burrow diameter. (ii) **Examine burrow wall**: is the burrow lined with mud or pellets? look for scratch marks; are laminae in adjacent sediment deflected by the burrow? (iii) **Examine burrow fill**: is it different from adjacent sediment? – coarser or finer grain-size; richer or poorer in skeletal debris; more or less iron (shown by colour, red/yellow/brown); is the fill pelleted? are there curved back-fill laminae within the burrow-fill sediments? Has the fill been dolomitised or silicified? Have nodules grown around or within the burrows? (iv) **Look for spreite**, i.e. curved laminae associated with vertical U-shaped burrows.

5.6.2   Trace fossils

Trace fossils are best considered in terms of their mode of formation; the principal groups are (i) locomotion (crawling, walking, running, etc.) tracks and trails; (ii) grazing trails; (iii) resting traces, occurring on bedding surfaces and undersurfaces; (iv) feeding burrows; and (v) dwelling burrows, mainly occurring within beds (Table 5.9 and Figure 5.70). All

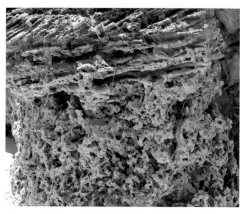

Figure 5.68 *Intensely bioturbated sediment. Field of view 0.5 m across. Outer shoreface grainstone. Pleistocene, Western Australia.*

Table 5.8 *Bioturbation index where each grade is described in terms of the sharpness of the primary sedimentary fabric, burrow abundance and amount of burrow overlap. The percentage bioturbated is just a guide and not an absolute division; see Figure 5.69 for a diagrammatic representation of the various grades.*

Grade	Percent bioturbated	Classification
1	1–5	Sparse bioturbation, bedding distinct, few discrete trace fossils and/or escape structures
2	5–20	Little bioturbation, bedding distinct, low trace-fossil density
3	20–50	Moderate bioturbation, bedding still visible, trace fossils discrete, overlap rare
4	50–80	Abundant bioturbation, bedding indistinct, high trace-fossil density with overlap common
5	80–95	Intense bioturbation, bedding completely disturbed (but just visible), later burrows discrete
6	95–100	Complete bioturbation, sediment reworking due to repeated overprinting

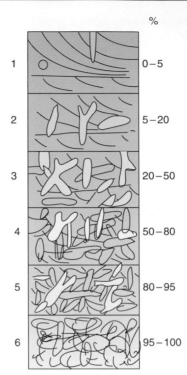

%

1   0–5

2   5–20

3   20–50

4   50–80

5   80–95

6   95–100

Figure 5.69 Different degrees and grades of bioturbation that generate the ichnofabric.

these structures are made by animals in unconsolidated sediments, clastics or carbonates. A further type of trace fossil is (vi) a *boring*, made by organisms into a hard substrate, a cemented sediment, pebble or fossil.

Locomotion tracks are produced by animals on the move, and so are usually straight or sinuous trails on bedding surfaces, contrasting with the more complicated feeding and grazing structures (see below). They can be produced by many types of animal in almost any environment. Common crawling traces are made by crustaceans, trilobites (e.g. *Cruziana*) and annelids. Vertebrates such as reptiles (especially the dinosaurs), amphibians and mammals leave *footprints* as trace fossils. See Figure 5.71 for features to measure if you find footprints.

Table 5.9 *Main features of trace fossil groups.*

Locomotion tracks: tracks, uncomplicated pattern; linear or sinuous, includes footprints

Grazing trails: more complicated surface trails, symmetrical or ordered pattern; coiled, radial, meandering, mostly made by detritus feeders

Resting traces: impression of where animal rested during life (but not a fossil mould)

Dwelling burrows: simple to complex burrow systems but no suggestion of systematic working of sediment; burrows can be clay lined or pelleted; some made by suspension feeders

Feeding burrows: simple to complex burrow systems; often well-organised and defined branching pattern indicating systematic reworking of sediment by detritus feeders

Borings: mostly simple structures (tubular, flask-shaped), cutting hard substrates such as pebbles, fossils and hardgrounds

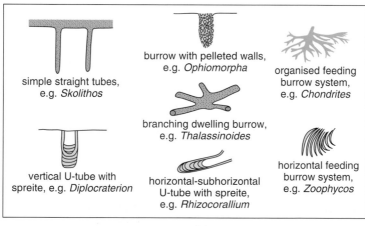

simple straight tubes, e.g. *Skolithos*

burrow with pelleted walls, e.g. *Ophiomorpha*

organised feeding burrow system, e.g. *Chondrites*

branching dwelling burrow, e.g. *Thalassinoides*

vertical U-tube with spreite, e.g. *Diplocraterion*

horizontal-subhorizontal U-tube with spreite, e.g. *Rhizocorallium*

horizontal feeding burrow system, e.g. *Zoophycos*

Figure 5.70 *Sketches of common dwelling and feeding burrows.*

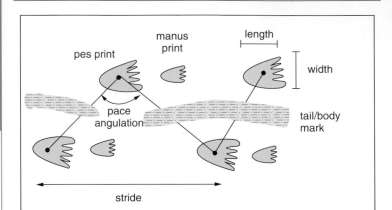

Figure 5.71 *Features to measure with footprints: width and length of each footprint, stride and pace angulation.*

Grazing traces occur on the sediment surface and are produced by deposit-feeding animals systematically working through the surface sediment for food. These biogenic structures usually consist of well-organised, coiled, meandering and radiating patterns. Grazing traces tend to occur in relatively quiet, often deeper-water, depositional environments and are produced by such organisms as molluscs and crustaceans. Examples are *Helminthoides*, *Palaeodictyon* and *Nereites* (e.g. Figure 5.72).

Resting traces are made by animals sitting on a sediment surface and leaving an impression of their body. Although quite rare, resting traces of starfish (e.g. *Asteriacites*) and bivalves (e.g. *Pelecypodichnus*) occur in shallow-water sediments.

Burrow structures can be very simple or quite complex. Features to note are shown in Figure 5.73. *Dwelling structures* (Figure 5.70) are burrows, varying from simple vertical tubes (e.g. *Skolithos*, Figure 5.74, *Monocraterion*) to U-shaped burrows, oriented either vertical (e.g. *Arenicolites*, *Diplocraterion*), subvertical or horizontal (e.g. *Rhizocorallium*) to the bedding. With U-shaped burrows, concave-up laminae, termed *spreite*, occur between and below the U-tube and form through the upward and downward movement of the animal in response to sedimentation and erosion. *Planolites* is a simple

Figure 5.72 *Trails on bedding plane. Field of view 20 cm across. Outer shoreface sandstone. Permian, Carnarvon, Western Australia.*

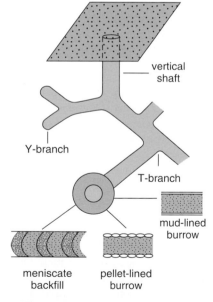

Figure 5.73 *Features of burrows.*

Figure 5.74 *Simple vertical dwelling burrows,* Skolithos. *Field of view 50 cm across. Shoreface cross-bedded sandstone. Permian, Dongara, Western Australia.*

unbranching, unlined burrow, probably a crustacean dwelling structure (Figure 5.75).

When burrowing organisms are rapidly buried, some can move up to regain their position relative to the sediment-water interface; in so doing they leave behind a characteristic *escape structure*, which deflects adjacent laminae to give a chevron structure. *Conichnus*, formed by a sea anemone, is of this type.

Other burrows, particularly those of crustaceans, are simple to irregularly branching systems, with walls composed of pellets or clay (e.g. *Ophiomorpha* and *Thalassinoides*; see Figure 5.70). Some dwelling burrows possess curved laminae (meniscate) (Figure 5.70), indicating back-filling by the animal as it moved through the sediment (e.g. *Beaconites*, a lungfish burrow found in fluvial sediments).

Feeding structures are trace fossils developed within the sediment by deposit-feeding organisms searching for food. One of the commonest is a simple, non-branching, back-filled horizontal to subhorizontal burrow (diameter 5–20 mm); some are composed of a series of sand balls, giving a beaded appearance (e.g. *Eione*, Figure 5.76). Other feeding burrows are highly organised with regular branching patterns

Figure 5.75 Simple burrows, unbranching and unlined, Planolites. *Field of view 20 cm across. Mid-shelf bioclastic wackestone. Carboniferous, NE England.*

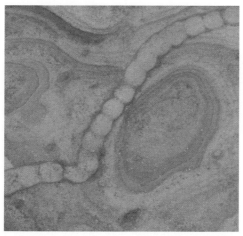

Figure 5.76 Beaded feeding burrow, Eione, *the result of the animal ingesting and excreting sand (to remove the organic matter) as it burrowed through the sediment. Field of view 20 cm across. Outer shoreface lithic arenite. Carboniferous, NE England.*

Figure 5.77 *Feeding burrow,* Zoophycos. *Field of view 20 cm across. Mid-shelf bioclastic wackestone. Carboniferous, NE England.*

Figure 5.78 *Feeding burrow from deeper-water facies,* Spirorhaphe. *Field of view 30 cm across. Pelagic lime mudstone. Tertiary, Paxos, Greece.*

(e.g. *Chondrites*, Figure 5.70); others have regular changes of direction, produced by the organism systematically working through the sediment extracting organic matter – e.g. *Zoophycos* (Figure 5.77) and *Spirorhaphe* (Figure 5.78).

Burrow structures within a bed may descend from the bedding plane above, which was the sediment surface when the burrowing organisms were active. In some beds there may be a number of different burrow structures, occurring at different depths relative to the original seafloor. This is referred to as a *tiering* of the trace fossils. Look for the presence of different burrows, for example, large ones and small ones, lined and

Figure 5.79 *Borings, 'flask-shaped', of Miocene lithophagid bivalves into grey Cretaceous dolomite at an unconformity surface, overlain by bioclastic packstone. Field of view 10 cm across. Tarragona, Spain.*

unlined, simple tubes and branching burrows, stuffed and simply filled, and for one type of burrow cross-cutting another.

Borings, traces made in solid rock or a fossil, vary in shape from tubular structures to oval/round holes, all generally filled by sediment. In some cases the host sediment has been weathered away to leave the borings standing proud. Borings made by lithophagid bivalves have a distinctive 'round-bottomed flask' shape (Figure 5.79) and the shell itself may still be present. Sponge borings tend to have a beaded appearance with scalloped margins, whereas annelid (polychaete) borings are more uniform bores.

Borings are common in hardground surfaces (see Figures 5.46 and 5.47, and Section 5.4.3) and are important evidence that the seafloor was cemented during deposition. Look for the borings cutting shells within the limestone and penetrating shells encrusting the hardground surface itself. Pebbles and intraclasts in shallow-marine sediments are commonly bored as well as many larger fossils, such as corals, echinoids and oysters. Unconformity surfaces may also be bored (e.g. Figure 5.79).

5.6.3 Use of trace fossils in sedimentary studies
Biogenic sedimentary structures can give vital information for environmental interpretation, in terms of water depth, salinity, energy

level, oxygenation, and so on, and they are especially valuable where body fossils are absent. It may be possible to recognise different suites or assemblages of trace fossils in a sedimentary succession so that *ichnofacies* can be distinguished. Four major marine ichnofacies are recognised, named after the typical trace fossil present, occurring in littoral (*Skolithos* ichnofacies), sublittoral (*Cruziana*), bathyal (*Zoophycos*) and abyssal (*Nereites*) zones. See Table 5.10 for further information on these trace fossil assemblages, their depositional environments and facies context.

Although particular trace fossils are typically found in a certain facies, similar conditions can exist in other environments, so that the same trace fossils occur there too. The trace fossils do not have to have been made by the same animals of course – just that they had the same behaviour. Thus, for example, ichnofossils typical of moderate-low energy, quiet, deepish water, such as those of the *Cruziana* and *Zoophycos* ichnofacies, may occur in shallow-water lagoonal facies too.

Trace fossils can give an indication of sedimentation rate. Intensely bioturbated horizons and beds with well-preserved, complex feeding and grazing traces would generally reflect slow rates of sedimentation. Bored surfaces (hardgrounds; see Section 5.4.3) generally indicate a break in sedimentation (a hiatus or omission surface) allowing seafloor cementation to take place. Intense bioturbation may occur immediately below a flooding surface, for example, where sandstones are sharply overlain by deeper-water mudrocks. U-shaped burrows with spreite and escape structures reflect more rapid sedimentation.

Trace fossils can give an indication of *sediment consistency*, and five states are recognised: soupground, softground, looseground, firmground and hardground (Table 5.11). If the sediment is soft, then tracks made on a lamina can be transmitted through to underlying laminae (Figure 5.80). Where burrows are either filled with coarser sediment or are preferentially lithified during early diagenesis, surrounding sediments are commonly compacted around the burrow fills if the sediment was a softground. In firmgrounds, the burrows show little compaction, and hardgrounds are recognised by the presence of borings and encrusting organisms.

Trace fossils seen on a bedding-plane may show a preferred orientation, reflecting contemporaneous currents (e.g. bivalve resting traces

Table 5.10 Trace fossil assemblages (ichnofacies), depositional environments, facies context, typical trace-fossil types and ichnofossil examples.

Skolithos assemblage	Cruziana assemblage	Zoophycos assemblage	Nereites assemblage
Sandy shoreline, foreshore and shoreface, 0–10 m depth, high energy	Sublittoral zone, offshore inner shelf, ~10–100 m depth, also lagoons	Bathyal zone, outer shelf, slope, shallow basin, ~100–2000 m depth, also lagoons	Bathyal/abyssal zone, deep-sea floor, ~1000–5000 m depth
Facies: flat-bedded or cross-bedded medium/coarse sandstones and grain/packstones	Facies: parallel- and cross-laminated fine sand/silt-stones, pack/wacke-stones and mudrocks	Facies: laminated fine sand/siltstones, mudrocks and wacke/mudstones	Facies: mudrocks and lime mudstones ± turbidites
Low diversity, vertical burrows, simple, U-shaped, pellet-lined	Wide variety of surface tracks and trails, complex burrows	Limited number of feeding + grazing trails and burrows, some complex	Regular patterns on sediment surface, often seen on bed undersurface
Skolithos, Ophiomorpha, Diplocraterion, Monocraterion	Cruziana, Asteriacites, Rhizocorallium, Eione, Chondrites, Planolites, Thalassinoides	Zoophycos, Phycosiphon, Spirophyton	Nereites, Helminthoides, Palaeodictyon, Spirorhaphe

5. Sedimentary Structures

6. Fossils in the Field

7. Palaeocurrent Analysis

8. Facies Identification

Table 5.11 *Types of substrate and their recognition from trace fossils.*

Sediment consistency	Nature of original sediment	Trace fossils and features
Soupground	Water-saturated, clay-rich sediment	Trace fossils highly compressed and smeared
Softground	Soft mud	Substantial compaction of burrows so indefinite outlines
Looseground	Loose sorted sand and silt	Burrows well-defined, often with linings or pelleted walls, some compaction
Firmground	Stiff substrate, e.g. muddy sand	Burrow outlines sharp and minor compaction
Hardground	Seafloor-cemented sediment, usually limestone	Borings cutting sediment and grains/fossils, encrusting fossils, clasts

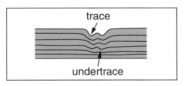

Figure 5.80 *Trace made by animal on sediment surface transferred to lower laminae to form an undertrace.*

commonly show this). Some trace fossils (footprints, U-shaped and escape burrows) can be used to show the way-up of strata.

See Recommended Reading for more information.

5.6.4 Rootlet beds

Root systems of plants disrupt the internal structure of beds in a similar way to burrowing animals. Many roots and rootlets are vertically arranged whereas others are horizontally disposed; many branch freely. Roots are commonly carbonised, appearing as black streaks; many are preserved as impressions, but some roots are preserved as casts, consisting of sandstone (e.g. *Stigmaria*). The identification of rootlet beds indicates in situ growth of plants and soil development, and thus subaerial conditions; coal seams may occur above such rootlet beds (see Section 3.10).

Plant debris is easily transported, however, so that plant-rich sediments are common. Examine a plant bed for rootlets; if it is just a collection of plant debris washed in, much of the material will be of the subaerial parts of the plant: leaves and branches. Sea grass, growing in several metres of water, common today, evolved in the Tertiary; their rootlets would not indicate subaerial exposure.

Where plants have been growing in a semi-arid environment, calcretes may develop within the soil profile and the plant roots may then become calcified or coated in calcrete to form *rhizocretions* (see Figure 5.63 and Section 5.5.6.2).

5.7 The Geometry of Sedimentary Deposits and Lateral Facies Changes

Some sedimentary rock units can be traced over large areas and show little change in character (i.e. facies) or thickness; others are laterally impersistent. The geometry of sedimentary deposits should be considered on the scale of the individual bed or rock unit as seen in an exposure, and on a larger, more regional scale, in terms of the shape of the sediment body or package of a particular lithofacies or group of related lithofacies.

5.7.1 Sediment body geometries

The geometry of an individual bed or rock unit can be described as being *tabular* if laterally extensive, *wedge-shaped* if impersistent but with planar bounding surfaces, and *lenticular* if one or both of the bounding surfaces is curved (Figure 5.81). For the larger scale, the term *sheet* (or blanket) is often applied if the length-to-width ratio of the sedimentary unit is around 1:1 and the sediment body covers a few to thousands of square kilometres. Elongate sediment bodies, where the length greatly exceeds the width, can be described as *linear* (also ribbon or shoestring) if unbranching, *dendroid* if branching, and a *belt* if composite. Many elongate sand bodies are channel fills, oriented down the palaeoslope. Elongate sediment bodies can also form parallel to

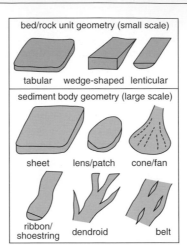

Figure 5.81 *Common geometries of beds or rock units (on a scale of metres to tens of metres) and sediment bodies (on kilometre or regional scale).*

a shoreline, as in beach and barrier island developments. Sedimentary units can be discrete entities forming *lenses* and *patches*, the latter term being particularly applicable to some reef limestones. Coarse clastic sediments may form fans, wedges and aprons where deposited at the toe of a slope. Examples include alluvial fan, fan-delta, toe-of-slope and submarine fan deposits.

Facies, defined by the lithological, textural, structural and palaeontological features of the sedimentary rock, commonly change laterally as well as vertically in a sedimentary succession. This can involve a change in one or all of the parameters defining the facies. Lateral changes can be very rapid, over several or tens of metres, or more gradational, when the change takes place over several kilometres. Facies changes reflect changes in the environmental conditions of sedimentation (see Chapter 8).

Observations of the geometry of individual beds and rock units normally present no problem. In a quarry or cliff exposure, follow beds laterally to determine their shape and then make notes and sketches (or take photographs). With the larger-scale geometry of sediment bodies, if exposures are very good as in some mountainous and vegetation-free regions, it may be possible to see the lateral changes and sediment-body shape directly, and to see the relationships between sedimentary units. Where exposure is limited, make detailed logs across the same part of the rock succession at several or many localities. To be sure that sections are equivalent, it is necessary to have either a laterally continuous horizon in the succession (a marker bed), or the presence of the same zonal fossils. If lateral facies changes are suspected in an area of poor exposure then detailed mapping and logging of all available exposures may be required to demonstrate such changes.

5.7.2   Stratigraphic relationships: lapouts and truncation

From the study of seismic sections, several types of large-scale relationship between sediment bodies, loosely referred to as lapouts, have been noted: onlap, downlap, offlap and toplap, as well as truncation (Figure 5.82). It is important to be able to document such relationships in a succession, since they relate to changes in relative sea level and *accommodation space* (concepts used in sequence stratigraphy; see Sections 2.8 and 8.4). However, in the field, they are rarely seen unless large coastal, mountainside or quarry exposures are available; they may be revealed by regional stratigraphic studies and mapping.

Downlap may be seen as gently to steeply dipping surfaces (the clinoforms described in Section 5.3.3.14, for example, are downlapping beds; see Figures 5.36, 5.84 and 5.86), and these may pass basinwards to a prominent surface. *Offlap* is where sediments are building out into the basin (Figure 5.84). Downlap-offlap represents progradation of a sedimentary unit through normal or forced regression, and the *downlap surface* itself is usually a thin, condensed bed or sediment-starved

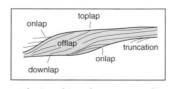

Figure 5.82 *Lapout relationships between sedimentary units: onlap, offlap, toplap, downlap and truncation.*

Figure 5.83 *Massive reef limestone (microbial boundstone) to left with offlapping reef-debris beds downlapping to the right onto condensed deeper-water lime mudstone. Cliff is 40 m high. Triassic, Catalonia, Spain.*

Figure 5.84 *Prograding, offlapping oolitic grainstone with gentle clinoform slope. Notice that several packages can be discerned, separated by prominent mudstone layers. Quarry face 40 m high. Jurassic, Yorkshire, NE England.*

horizon with much bioturbation, perhaps with glauconite or phosphorite. This was an area of little or no sedimentation, until buried by the advancing clinoform/slope sediments. In some instances with clinoforms, *toplap* may be visible, where the sloping bed feathers out landwards (typical of normal regression). In other cases there is truncation at the top of the offlapping beds (see below) as a result of subsequent erosion (often through forced regression and base-level fall).

Figure 5.85 *Onlap of sedimentary strata. Notice the subtle discontinuity within this package of shallow-water limestones and mudrocks, as a result of slight tilting of strata below the unconformity. Height of cliff 80 m. Jurassic, Mareb, Yemen.*

Onlap of one sedimentary unit onto or against another is a common large-scale arrangement but again it is rarely observed in one exposure. It can be seen where strata gradually bury a unit with topographic expression, a carbonate platform margin or reef, for example, or where strata fill a large, broad channel structure and lap up onto its margin, or where strata onlap a tilted surface (Figure 5.85). However, onlap on a larger scale can be demonstrated by determining the age of the base of a stratigraphic unit over a large area, when it may then be apparent that it youngs in a particular direction, that is, it is onlapping the underlying strata that way. The nature of the onlap surface itself (which may be an *unconformity*) may also change laterally; it may become a more prominent palaeokarst or palaeosoil, for example in the direction of onlap, as a result of a longer time available for its development. Onlap of this type in marine strata reflects a rise of relative sea level (transgression) and the onlap surface may be a sequence boundary and/or transgressive surface (see Section 8.4).

A *truncation surface* is where strata are cut off beneath a prominent bedding plane. Again this may be visible in good outcrops but commonly it is on a larger scale, and examining exposures over a large area may be necessary to identify this type of stratigraphic boundary (an *unconformity*). Truncation surfaces are the result of uplift and tilting, perhaps folding and erosion of strata. Truncated surfaces may also be onlap surfaces, with the age of the sediments immediately beneath the surface getting older in the direction of the younging of the overlying, onlapping unit. This would reflect the increased time for erosion of the rocks below the onlapping unit (see Figure 5.85 for an example).

For these larger-scale relationships between stratigraphic units, look carefully at the bedding planes in a good and extensive outcrop and follow them along.

- Is there any gentle dip of beds down to a surface? This would be *downlap* (downlap may involve more steeply dipping beds but this should be obvious).
- Is there any truncation of underlying strata up to a prominent bedding plane? This would indicate an *unconformity*.
- Laterally, do beds overlap the one below (which may be dipping slightly), onto a prominent surface? This would be *onlap*.

These may all be very subtle arrangements so that you may need to study the exposure carefully, perhaps by standing back or by using binoculars on cliff-faces, to see these features. Bear in mind that these angular relationships between stratigraphic units do depend on the plane of the section, and with some outcrops there is the problem of perspective, looking up to high cliffs, for example.

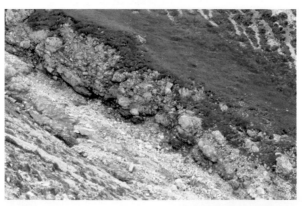

Figure 5.86 *Large-scale clinoform bed, cliff 50 metres high, composed of metre-scale blocks of shelf-margin limestone. Triassic, Dolomites, Italy.*

6

FOSSILS IN THE FIELD

6.1 Introduction

Fossils are an important component of sedimentary rocks. First and foremost, they can be used for *biostratigraphic* purposes to determine the relative age of the rock succession and for correlations with successions elsewhere. The identification of fossils to the species level is not easy and in most cases it is best left to the specialist. However, there are various fossil handbooks published and the *Treatise of Invertebrate Paleontology* that can be used to identify fossils.

Fossils are of great use in the environmental interpretation of sedimentary rocks, and in this context many useful observations can be made in the field by the non-specialist with a keen eye. Fossils can tell you about the water-depth, level of turbulence, salinity and sedimentation rate; they can record palaeocurrent directions and give information on palaeoclimatology. In some cases the whole environmental interpretation of a sedimentary succession may depend on the presence of just a few fossils, and occasionally an interpretation has been shattered by the discovery of new fossils. Field observations on fossils should consider their distribution, preservation (taphonomy) and relation to the sediment, their associations and diversity.

A checklist for the examination of fossils in the field is given in Table 6.1, and the distribution, diversity and abundance of the main fossil groups through the Phanerozoic is shown in Figure 6.1. In Precambrian strata only microbialites, especially stromatolites, are seen in the field.

6.1.1 Macrofossils

Macrofossils are big enough to see in the field, and with a little experience or a course in palaeontology it is relatively easy to identify the animal group to which they belong, and in some cases even a species

Sedimentary Rocks in the Field: A Practical Guide, Fourth Edition   Maurice E. Tucker
© 2011 John Wiley & Sons, Ltd

Table 6.1 *Checklist for the examination of fossils in the field: (A) distribution of fossils in sediment, (B) fossil assemblages and diversity, and (C) diagenesis of fossil skeletons.*

(A) Distribution of fossils in sediment

1. Fossils largely in growth position

(a) Do they constitute a reef? Characterised by: colonial organisms; interaction between organisms (such as encrusting growth); presence of original cavities (filled with sediment and/or cement) and massive, unbedded appearance (see Section 3.5.3). (i) Describe growth forms of colonial organisms (see Figure 3.13); do these change up through the section/reef? (ii) Are some skeletons providing a framework?

(b) If non-reef, are fossils epifaunal or infaunal? If epifaunal, how have fossils been preserved (e.g. by smothering)?

(c) Do epifaunal fossils have a preferred orientation, reflecting contemporary currents? If so, measure orientation.

(d) Are fossils encrusting substrate, i.e. is it a hardground surface?

(e) Are the plant remains rootlets?

2. Fossils not in growth position

(a) Are they concentrated into lenses, pavements or stringers or are they laterally persistent beds or evenly distributed throughout the sediment?

(b) Do fossils occur in a particular lithofacies? Are there differences in the faunal content of different lithofacies?

(c) If fossil concentrations occur, what proportion of fossils are broken and disarticulated? Are delicate skeletal structures preserved, such as spines on shells? Check sorting of fossils, degree of rounding; look for imbrication, graded bedding, cross-bedding, scoured bases and sole structures.

(d) Do fossils show a preferred orientation? If so, measure orientation.

(e) Have fossils been bored or encrusted?

(f) Note degree of bioturbation, ichnofabric and any specific trace fossils present.

Table 6.1 *(continued)*

(B) Fossil assemblages and diversity
1. Determine the composition of the fossil assemblage by estimating the relative abundance of the different fossil groups in a bed or on a bedding plane.
2. Is the fossil assemblage identical in all beds of the section or are there several different assemblages present? If the latter, do the assemblages correlate with different lithofacies?
3. Consider the degree of reworking and transportation; does the fossil assemblage reflect the community of organisms that lived in that area?
4. Consider the composition of the fossil assemblage. For example, is it dominated by only a few species? Are they euryhaline or stenohaline? Are certain fossil groups conspicuous by their absence? Do all fossil groups present have a similar mode of life? Do pelagic forms dominate? Are infaunal organisms absent?

(C) Diagenesis of fossil skeletons
1. Is original mineralogy preserved or have skeletons been replaced – calcified, dolomitised, silicified, hematised, pyritised, etc.?
2. Have fossils been dissolved out to leave moulds?
3. Do fossils occur preferentially in nodules?
4. Are fossils full-bodied or have they been compacted?

name can be applied. The common fossil types you should be able to recognise easily in marine Palaeozoic strata are the trilobites, graptolites, brachiopods, gastropods, ammonoids, orthocones, rugose corals, tabulate corals, bryozoans, stromatoporoids and crinoids. In Mesozoic strata the common marine fossils are brachiopods, bivalves (including the rudists in the Cretaceous), gastropods, scleractinian corals, crinoids, echinoids (especially in the Cretaceous), ammonites, belemnites and calcareous algae. In Cainozoic strata, you should find abundant bivalves and gastropods, along with scleractinian corals and calcareous algae. Other macrofossil groups, such as vertebrates and crustaceans, are rarely found. Plant fossils are abundant in certain strata, mostly non-marine facies of course.

SEDIMENTARY ROCKS IN THE FIELD

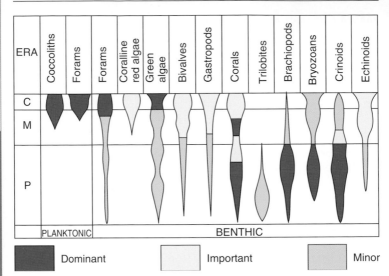

Figure 6.1 *The distribution, abundance and diversity (shown by width) of the main fossil groups through the Phanerozoic. C, Cainozoic; M, Mesozoic; P, Palaeozoic.*

6.1.2 Microfossils

These are not usually seen in hand specimens, except in rare cases as grains under a hand-lens (e.g. foraminifera and radiolarians), or in particular horizons where large forms of microfossil have developed, for example the nummulite forams in Eocene rocks. A sample of 0.5–1 kg of material collected in the field is usually sufficient for microfossil extraction back in the lab. Microfossils can be studied in thin-sections of the rock, but many need to be extracted whole and viewed with a binocular microscope or scanning electron microscope (SEM). Conodonts in Palaeozoic strata are obtained by dissolving limestones in dilute acid, and spores in post-Devonian strata are extracted with HF (hydrofluoric acid, but beware health and safety issues!). Foraminifera, ostracods, etc. can be removed from weakly cemented chalk and mudrock by crushing, sieving and ultrasound or repeated boiling and freezing. See micropalaeontological texts for techniques.

Microfossils can be particularly useful for biostratigraphic correlation – which ones depend on the age and facies. They can also be used in palaeoenvironmental studies, to separate marine from non-marine facies, for example, or hypersaline from hyposaline, or shallow from deep.

6.2 Fossil Distribution and Occurrence

Pay attention to the distribution of the fossils while studying or logging a succession of sediments. Fossils can be evenly distributed throughout a rock unit without any preferential concentration in certain beds; this generally only occurs where the sediment is homogeneous throughout. Use the percentage charts in Figure 3.3 to estimate the proportions of the various fossils present.

In many cases, fossils are not evenly distributed but occur preferentially in certain beds, lenses or in a buildup (a reef). The term *coquina* (and lumachelle) is often used for an accumulation of shells. Always examine the types of fossil present and determine their relative distribution. See if there is a correlation between fossil type and lithofacies. Concentrations of fossils at particular levels can arise from current activity or preferential growth through favourable environmental conditions.

Fossils can be arranged in several different ways within a bed: concordant, oblique, perpendicular, imbricated, stacked and nested (see Figures 6.2 and 6.3 for examples). Examine the fossiliferous bed in cross-section and determine the orientation of the fossils; this depends on the levels of current activity and sorting (see below).

On the bedding planes, fossils may be arranged as a carpet or pavement a few valves thick – all lying flat upon the surface, or they may form a linear feature on the surface, a stringer or low ridge (Figure 6.2), concentrated there by currents.

6.2.1 In situ fossil accumulations

Where favourable conditions existed, some fossils will be preserved intact, with little breakage or disarticulation, and at least some of the organisms that lived on or in the sediment will be in their life position. Fossils that are commonly found in growth position include brachiopods (Figure 6.4), some bivalves (especially the rudists), corals (Figure 6.5), bryozoans and stromatoporoids. It is worth noting the general level of infaunal activity (the amount of bioturbation and the *ichnofabric*; see Section 5.6.1 and Figure 5.69) in conjunction with the body fossils present.

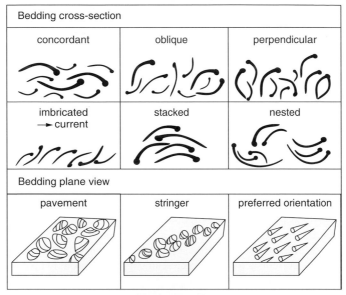

Figure 6.2 *The various arrangements of fossils in sedimentary rocks as seen on bedding planes and in bed cross-sections.*

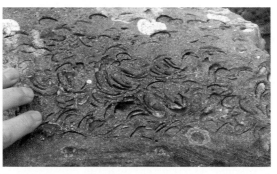

Figure 6.3 *Disarticulated shells of bivalves in a storm bed. Note the arrangement of the shells up through the bed: convex-up, imbricated, nested and convex-up, reflecting the varying effects of the current. Bed is 10 cm thick. Shoreface bioclastic grainstone. Pleistocene, Western Australia.*

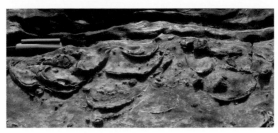

Figure 6.4 *Brachiopods (productids) in growth position (concave-up). Field of view 24 × 10 cm. Mid-shelf wackestone-packstone. Mid-Carboniferous, NE England.*

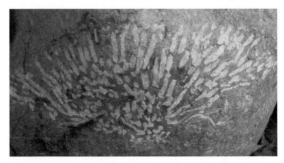

Figure 6.5 *Colonial coral in growth position. Also overturned brachiopod shell (convex-up). Field of view 15 × 10 cm. Mid-shelf wackestone-packstone. Lower Carboniferous, NE England.*

Fossils can be organised into *reefs* (the more general term is *buildups*) (see Section 3.5.3). In these the majority of the fossils are in their growth position, perhaps with some organisms growing over and upon each other. Colonial organisms can dominate and the rock characteristically has a massive, unbedded appearance (see Figures 3.11, 3.12 and 5.83). The fossils commonly have a variety of growth forms (shapes), depending on the environmental conditions (Figure 3.13), and there may be an upward change through the reef. Cavities, both large and small, perhaps filled with internal sediment and calcite cement, are common in reefal limestones (see, e.g., Figures 5.41 and 5.44).

199

Hiatal concentrations are fossil accumulations as a result of minimal deposition of mud and sand. The fossils may be bored, encrusted, discoloured or phosphatised, and authigenic minerals such as chamosite and glauconite may be present. These beds are usually highly condensed and biozones may be missing.

6.2.2   Current accumulations of fossils

Concentrations of skeletal material arising from current activity can form in a number of ways. Transportation of skeletal debris by storm currents leads to the deposition of *storm beds* (also called *tempestites*; see Section 5.3.3.9, Figure 5.29). These *event beds* tend to be quite laterally persistent, and possess sharp, usually scoured bases. Internally, they vary from laminated beds with normal size-grading and good sorting of fossils, to beds and lenses with no apparent sorting of fossils, the deposit having a 'dumped' appearance with a wide range of grain-sizes and no internal structure. Examine the arrangement of fossils in a storm bed by reference to Figure 6.2; Figure 6.3 is an example.

Fossil accumulations can also form through the winnowing action of weaker currents, which remove finer sediment and skeletal grains. Such *fossil lag deposits* are usually impersistent lenses; they too occur in shallow-shelf successions, especially at transgressive surfaces. Fossil-rich beds can also be formed through reworking of sediment by laterally migrating tidal channels.

The degree of current activity and reworking affects the proportion of broken and disarticulated carbonate skeletons within a fossil concentration. With an increasing level of agitation, fossils grade from perfect preservation, with all delicate structures intact and adjoined, to poor preservation where fossils are abraded and broken. Points to look for with specific fossils are:

- **crinoids:** the length of the stems and whether all ossicles are separated; if the calyx is present whether this is attached to the stem or not;
- **bivalved shells (bivalves, brachiopods and ostracods):** whether the valves are disarticulated or articulated; if the former, whether there are equal numbers of each valve, if the latter whether the valves

are open or closed; and also whether there is a preferred way-up of the valves or all orientations are present (see Figures 6.3 and 6.6);

- **some brachiopods and bivalves:** whether the spines for anchorage are still attached;
- **trilobites:** whether the exoskeleton is whole or incomplete (e.g. only tails).

6.2.3 Preferred orientations of fossils

Elongate shells or skeletons affected by currents commonly have a preferred orientation of their long axes (Figure 6.7). This alignment is generally parallel to the current, or less commonly normal to the current if the skeletal fragments were subject to rolling. Both orientations may occur (Figure 6.8). Preferred orientations are commonly found with crinoid stems, graptolites, cricoconarids (tentaculitids), elongate bivalve shells, turreted gastropods, solitary corals, belemnites, orthoconic nautiloids and plant fragments. Some fossil orientations may reflect current directions during life. In a collection of elongate fossils on a bedding plane, measure their orientation. Chapter 7 deals with palaeocurrent analysis.

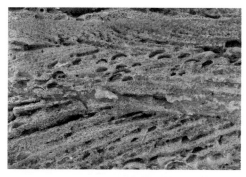

Figure 6.6 *Disarticulated shells of bivalves in a bidirectional (herring-bone) cross-bedded sandy limestone. Note the convex-up arrangement of most shells (this is the most stable position in a current), and that most shells have been dissolved out, since they were originally composed of the less stable mineral aragonite. Field of view 30 × 20 cm. Shoreface bioclastic grainstone. Pleistocene, Western Australia.*

Figure 6.7 *Graptolites showing a preferred orientation. Field of view 15 × 10 cm. Hemipelagic deepwater mustone. Ordovician, Wales.*

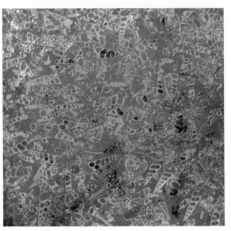

Figure 6.8 *High-spired gastropods showing two preferred orientations from the effects of a current. Field of view 10 × 10 cm. Bioclastic packstone. Tertiary, Tasmania, Australia.*

6.3 Fossil Associations and Diversity

6.3.1 Fossil assemblages

The fossil assemblages present, and their relationships to each other, give useful environmental information. First determine the fossil assemblage qualitatively by estimating the relative abundance of the different

202

fossil groups. For a precise analysis of the assemblage large blocks of the sediment need to be broken up carefully and all species identified and counted. This is best undertaken in the laboratory. If good bedding-plane exposures are available, count the number of each fossil species in a quadrat, a square of a particular area; $1\,m^2$ is usually taken. Plot the data in a histogram or pie diagram.

By carefully analysing the fossil assemblage from different beds in a section or different but coeval lithofacies over an area, changes in the assemblage can be recognised. Assemblages can be described by their dominant members; choose one or several characteristic forms present: for example, Productid-coral assemblage (common in Lower Carboniferous limestones), Echinoid-terebratulid-sponge assemblage (common in the Upper Cretaceous chalks).

A fossil assemblage is also a *death assemblage*. Many such assemblages are composed of the remains of animals that did not live in the same area. The skeletal debris would have been brought together by currents, and so generally consists of broken and disarticulated skeletons. However, some death assemblages consist of the skeletons of organisms that did live in the same general area. In these cases, some fossils may occur in their original life position and skeletal transportation would have been minimal. Reefs and other buildups are obvious examples of in situ death assemblages.

Where little transport of skeletal material has taken place after death, the fossil assemblage will reflect the community of organisms that lived in that area. A *community* can be referred to either by a dominant species, in the same way as an assemblage, or by reference to the lithofacies (e.g. a muddy sand community). The ascribing of an assemblage to a community is an important step, since a community is dependent on environmental factors, and changes in community can indicate changes in the environment. Once a community has been recognised, it is possible to look at the species present and deduce the roles the various organisms played in that community.

It must be remembered that much, if not most, of the fossil record is not preserved. There is obviously a bias towards the preservation of animal hard parts. In studying a community, thought should be given to animals and plants for which there is only circumstantial or no evidence. Trace fossils, pellets, coprolites and microbial laminae are important in this respect. In addition, there is growing evidence that

5. Sedimentary Structures

6. Fossils in the Field

7. Palaeocurrent Analysis

8. Facies Identification

at certain times in the Phanerozoic, organism skeletons composed of aragonite preferentially dissolved on the quite shallow seafloor, so that only calcitic fossils were then preserved.

One feature to look for in an assemblage is the presence of encrusting and boring organisms. Large skeletal fragments can act as substrates for others; oysters, bryozoans, barnacles, certain inarticulate brachiopods, algae and serpulid worms commonly encrust other skeletons (e.g. Figure 6.9). Boring organisms such as serpulids, lithophagid bivalves and sponges can attack skeletal fragments and other hard substrates, such as hardground surfaces (see Figure 5.47; Sections 5.4.2 and 5.6.2), pebbles and rock surfaces (as at unconformities; see Figure 5.79), producing characteristic holes and tubes. Boring and encrusting of skeletal debris tend to be more common where sedimentation rates are low. Burrows in sediment are also more common in such situations (see Section 5.6).

6.3.2 Fossils as environmental indicators

The number and type of species present in an assemblage depend on environmental factors. Where these factors (depth, salinity, agitation, substrate, oxygenation, etc.) are at an optimum, there is maximum species diversity. Infaunal and epifaunal benthic (bottom-dwelling), nektonic (free-swimming) and planktonic (free-floating) organisms are

Figure 6.9 Echinoid test encrusted with serpulid worm tubes, bryozoans and oyster. Field of view 7 × 5 cm. Upper Cretaceous, E England.

all present. Where there are environmental pressures, species diversity is lowered and certain aspects of the fauna and flora may be missing. However, in these situations, species that are present and can tolerate the environment may occur in great numbers. With increasing depth, pelagic fossils will dominate, such as fish, graptolites, cephalopods, planktonic foraminifera, posidonid bivalves and some ostracods. The same situation is found with an increasing degree of stagnation: benthic organisms will eventually be excluded altogether, so that only pelagic organisms are present.

6.3.2.1 Fossils and salinity

With salinities higher or lower than normal marine concentrations, many species are excluded completely. Groups tolerant of normal-marine conditions only (stenohaline forms) are corals, bryozoans, stromatoporoids and trilobites; many specific genera and species of other groups are also stenohaline. Some fossil groups (euryhaline forms) are able to tolerate extremes of salinity, for example certain bivalves, gastropods, ostracods and charophyte algae. Where sediments contain large numbers of only a few species of such groups, then hypersaline or hyposaline conditions should be suspected. In some cases, organism skeletons show a change in shape or size with extremes of salinity. Hypersaline conditions may well be indicated by the presence of evaporite pseudomorphs (see Section 3.6).

6.3.2.2 Fossils and depth

Depth of deposition is best inferred from sedimentary structures and facies; fossils can give an indication, but rarely anything precise. Many benthic fossils (brachiopods, bivalves, gastropods, corals, etc.) are typical of nearshore, agitated shallow-water environments, depths less than 20 m, say, although they may occur in deeper water too, but then generally in smaller numbers. Other fossils are more typical of the quieter, muddy shelf environment. As water depth increases, benthic fossils decrease in abundance and pelagic fossils become more common. The photic zone, around 100–200 m (depends on water clarity), could be recognised by the disappearance of algae, but there are actually few algal species living near this depth anyway.

In deeper water, the aragonite compensation depth (ACD) is the next threshold, at a few hundred to 2000 metres, below which aragonite shells

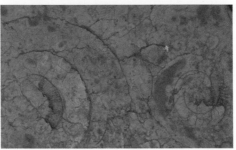

Figure 6.10 *Ammonites with no shell preserved since deposition took place below the aragonite compensation depth. Field of view 30 × 20 cm. Pelagic lime mudstone. Ammonitico Rosso, Jurassic, Italy.*

are not preserved, and the fossils are only casts. Ammonites in pelagic limestones can be preserved this way (Figure 6.10). The calcite compensation depth (CCD) is at several thousand metres, and below this no calcareous fossils are found; siliceous shales and cherts will be the typical sediments with only non-calcareous fossils (e.g. radiolarians and phosphatic ammonite aptychi).

6.3.2.3 Fossils (and trace fossils) and oxygen

The amount of oxygen in seawater and sediment porewater is an important factor controlling the organisms present, their preservation and the sedimentary facies. Five classes of oxygen-related facies are distinguished, with increasing oxygen content – anaerobic, quasi-anaerobic, exaerobic, dysaerobic and aerobic – and they have different trace fossils, body fossils and sediment textures and composition (Table 6.2). The degree of oxygenation is determined by water circulation, organic matter and sediment supply rates, and water depth.

The typical fossils and their occurrence in the major marine facies are shown in Table 6.3.

6.4 Skeletal Preservation (Taphonomy) and Diagenesis

The original composition of fossil skeletons is commonly altered during diagenesis. Many carbonate fossil skeletons were composed of aragonite when the animal was living. With the majority of such fossils and in

Table 6.2 *Fossils, ichnofabrics/trace fossils and oxygen: oxygen-related facies.*

Oxygen-related facies	Body fossils	Ichnofabric	Sediment
Anaerobic	No benthic fossils, well-preserved pelagic fossils	No trails or burrows, faecal material	Well-laminated, organic-rich, black muds
Quasi-anaerobic	Few microbenthic fossils, well-preserved pelagic fossils	Micro-bioturbation faecal material	Well-laminated, organic-rich, black muds
Exaerobic	Some in situ epibenthic macrofossils, small size, low diversity	Few shallow burrows, some trails	Laminated dark-grey muds/sands
Dysaerobic	Low diversity, small macrobenthics, thin shells	More shallow burrows, some deeper trails	Bioturbated laminae and beds of grey mud/sand
Aerobic	Diverse, large macrobenthics, heavy shells	Trace fossils may be abundant and diverse, tiering	Bioturbated strata, plus other sedimentary structures: ripples, cross-bedding, etc.

5. Sedimentary Structures

6. Fossils in the Field

7. Palaeocurrent Analysis

8. Facies Identification

Table 6.3 The main fossil components and preservation in marine facies.

Facies	Lithologies	Fossils	Diversity	Abundance	Taphonomy
Lagoonal facies, behind barrier or reef	Fine-grained clastics and carbonates, some coarser sediments from shelf by storms	Bivalves, gastropods, ostracods, trace fossils common	Low	Variable, but may be high	In situ faunas, some shell beds, introduced fossils from shelf by storms
Shoreline, beach and shoreface	Sandstones/grainstones + cross-bedding, SCS, HCS	Brachiopods, bivalves, gastropods, burrows	Low	Mostly low	In situ faunas rare, skeletal debris
Shoreface/inner-mid shelf	Sandstones/grainstones in storm beds, mudrocks	Brachiopods, bivalves, crinoids, echinoids, corals, many burrows	Variable	Moderate to high	Coquinas, shell-lags, in situ faunas rare, skeletal debris

Outer shelf	Mudrock mostly, thin storm beds	Brachiopods, trilobites, bivalves, graptolites, crinoids, ammonoids	Variable	Low	In situ faunas, rare lags or coquinas
Carbonate shelf margin	Reef limestones	Corals, bryozoans, stromatoporoids, molluscs, sponges, brachiopods, algae	High	High	In situ faunas, skeletal debris, coquinas
Slope and basin	Mudrocks, turbidite–debrite sandstones and limestones	Pelagic faunas (ammonoids, forams, etc.); shelf fossils in resedimented beds	Low	Variable, can be high	Pelagic and in situ faunas, skeletal debris

the normal course of events, the aragonite has been replaced by calcite. Other minerals that replace fossils include dolomite, pyrite, hematite and silica. In some cases, fossils can be dissolved out completely so that only moulds are left; this can happen preferentially to fossils originally composed of aragonite in a limestone or sandstone (see, e.g., Figures 6.3 and 6.6). During dolomitisation of a limestone, some fossils, notably those composed of calcite, are more resistant and may be left intact.

In rare cases, bioclasts are dissolved on the seafloor. This has usually occurred with aragonitic fossils in deeper-water environments where deposition took place below the ACD but above the CCD (see previous section and Figure 6.10). The casts of the fossils may then be encrusted by calcitic fossils or ferromanganese crusts.

Where early diagenetic nodules occur in a mudrock, fossils present within the nodules will be better preserved, that is, less compacted, compared with fossils in the surrounding mudrock (see Section 5.5.6). Nodule nucleation can take place preferentially around fossils; for example, the decay of a fish in sediment can set up a chemical microenvironment conducive for mineral precipitation. Ammonites occurring in mudrocks are commonly enclosed in calcareous nodules (Figure 6.11).

Figure 6.11 *Ammonite in a calcareous nodule: notice that the ammonite is not compacted but full bodied. The shell has been partly replaced by pyrite (indicating anoxic conditions within the sediment) and there are white calcite crystals filling the shell. Jurassic, NE England.*

7

PALAEOCURRENT ANALYSIS

7.1 Introduction

The measurement of palaeocurrents is a vital part of the study of sedimentary rocks, since they provide information on the palaeogeography, palaeoslope, current and wind directions and they are useful in facies interpretation. The measurement of palaeocurrents in the field should become a routine procedure; a palaeocurrent direction is an important attribute of a lithofacies and necessary for its complete description.

Many different features of a sedimentary rock can be used as palaeocurrent indicators. Some structures record the direction of movement (azimuth) of the current whereas others only record the line of movement (trend). Of the sedimentary structures, the most useful are cross-bedding and sole structures (flute and groove casts), but other structures can also give reliable results.

7.2 Palaeocurrent Measurements

The more measurements you can take from either a bed or a series of beds, the more accurate is the palaeocurrent direction you obtain, although an important consideration is the variability (spread) of the measurements.

First, assess the outcrop. If only one lithofacies is present measurements can be collected from any or all of the beds. If measurements from one bed (or many beds at an outcrop) are all similar (a unimodal palaeocurrent pattern; Figure 7.1), there is little point in taking a large number of readings. Some 20–30 measurements from an exposure would be sufficient to give an accurate vector mean (see Section 7.4). You should then find other outcrops of the same lithofacies in the immediate vicinity and farther afield, so that the palaeocurrent pattern over the area can be deduced.

Sedimentary Rocks in the Field: A Practical Guide, Fourth Edition   Maurice E. Tucker
© 2011 John Wiley & Sons, Ltd

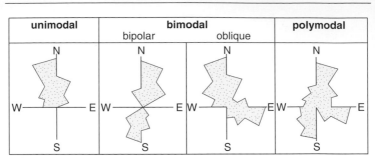

Figure 7.1 *The four types of palaeocurrent pattern, plotted as rose diagrams (with 30° intervals). The convention is to plot palaeocurrent azimuthal data in a 'current-to' sense, so that, for example in the unimodal case above, the current was flowing from the south towards the north.*

If readings vary considerably within a bed, you will need to collect a large number (more than 20 or more than 50 depending on the variability) to ascertain the mean direction.

Measurements collected from different sedimentary structures should be kept apart, at least initially. If they are very similar they can be combined. Also keep separate measurements from different lithofacies at an exposure; they may have been deposited by different types of current, or currents from different directions. Readings should be tabulated in your field notebook (Figure 7.2).

The measurements you have taken may not represent the current direction if either the shape or the orientation (or both) of the sedimentary structures has been changed by tectonism. It is important to appreciate that two changes can occur: tilt and deformation. A simple change in the inclination of the plane, of which the sedimentary structure is a part, is described as the tilt. Tilt does not change the shape of a sedimentary structure. Processes that change the shape of a sedimentary structure are described as deformation.

7.2.1 Correction of measurements for tectonic tilt

To ascertain the direction of palaeocurrents from structures in dipping beds it is necessary to remove the effects of tilting, a straightforward

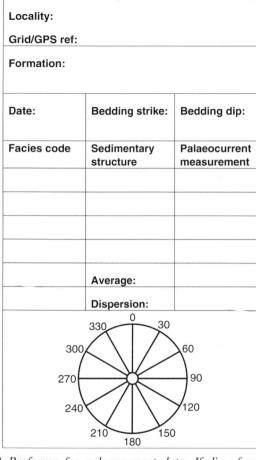

Locality:

Grid/GPS ref:

Formation:

Date:	Bedding strike:	Bedding dip:
Facies code	Sedimentary structure	Palaeocurrent measurement
	Average:	
	Dispersion:	

Figure 7.2 *Proforma for palaeocurrent data. If dip of cross-beds is required, add another column. Corrections for tectonic tilt are required for linear structures such as groove marks if the strata are dipping more than 25°, and for planar structures such as cross-bedding if the strata are dipping more than 10°. If this is the case, then add columns for the pitch and down-dip trend of linear structures or the dip and strike of cross-beds. The proforma sheet can be downloaded from the Internet at: www.wiley.com/go/sedimentaryrocks4e.*

5. Sedimentary Structures

6. Fossils in the Field

7. Palaeocurrent Analysis

8. Facies Identification

process that is described below. To do the same with deformed sedimentary structures is not simple and requires an accurate assessment of the strain the rock mass containing the sedimentary structure has undergone; description of how this is done is beyond the scope of this book but accounts can be found in structural geology textbooks. The tell-tale signs of deformed rock masses are the presence of such features as cleavage, minor folding, metamorphic fabrics and deformed fossils.

7.2.1.1 Correction of measurements for tectonic tilt: linear structures

The following steps should be taken to correct the trend of a *linear structure*, such as a flute or groove cast or parting lineation, that has been changed by a tilt of more than 25° (also see Figure 7.3).

1. First measure the direction-of-dip (or strike) and the angle-of-dip of the bedding surface (or undersurface); then measure the acute angle between the elongation of the structure and the strike of the bedding (this is the pitch or rake of the structure) and note the down-dip direction of the structure.
2. With the angle-of-dip and the direction-of-dip of the bed, plot the bedding surface as a great circle using the stereonet inside the back cover of this book (also available on the Internet at: www.wiley.com/go/sedimentaryrocks4e). To do this, place tracing paper on the net, draw the circle and mark north, south, east, west and the central point; mark the direction-of-dip on the circle (circumference of the net) and then rotate the paper so this mark is on the east-west (equatorial) line of the net; count in the angle-of-dip of the bed from the mark on the circumference and draw the great circle using the line already there on the net.
3. On the great circle, from the appropriate end, mark the acute angle between the trend of the sedimentary structure and the strike of the bedding surface (i.e. the pitch). To do this, have the paper in the same position as when you drew the great circle (dip direction on the net equatorial line) and then count the pitch angle from the end of the great circle to which the structure is directed down-dip.
4. From this point of the pitch angle on the great circle, move along the small circle to the nearest point on the circumference of the net and mark this place.

5. Finally rotate the tracing paper back to its original position with regard to the stereonet (north on north) and read off the new direction for the structure. For asymmetric linear structures such as flute marks the azimuth is obtained, whereas with structures giving a trend such as groove marks and current lineation, the current could have been in the opposite direction too.

See Figure 7.3 for a worked example.

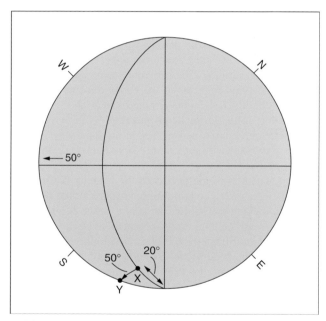

Figure 7.3 *An example of correcting the orientation of a linear structure for tectonic tilt. A bed dips at 50° to 225° and has a linear structure, such as a groove mark or parting lineation, with a pitch of 20° to the SE. The bed is plotted as a great circle on a stereonet and the pitch of the structure is marked on that circle (X). The bed is restored to the horizontal and with it the structure along a small circle to give its original orientation at point Y. On rotating back, the azimuth of Y is given: 155°. See text for details.*

7.2.1.2 Correction of measurements for tectonic tilt: planar structures, especially cross-bedding

The following steps should be taken to determine the original orientation and angle-of-dip of a planar structure (principally cross-bedding) that has been changed by a tilt of more than $10°$:

1. First measure the direction-of-dip (or strike) and the angle-of-dip of the bedding surface and then the direction-of-dip (or strike) and angle-of-dip of the cross-bedding.
2. Plot the pole for the bedding surface using the stereonet inside the back cover of this book. To do this, place tracing paper on the net, draw the circle and mark north, south, east, west and the central point. Mark the point of the direction-of-dip on the circle (circumference of the net) and then rotate the paper so this point is on the east-west/equatorial line of the net; count the angle-of-dip of the bed from the net central point along the equatorial line *away* from the point of the dip direction on the circumference. This point is the pole to the bedding on the lower hemisphere.
3. Plot the pole for the cross-bed surface in the same way as the pole to the bedding (so start with the tracing paper back at north, coinciding with the stereonet).
4. Restore the bedding to the horizontal by rotating the tracing paper to bring the pole of the bedding to the equatorial line of the net and (notionally) moving the pole to the centre of the net; then move the pole of the cross-bedding along the small circle it is now situated upon, in a similar direction, by the same number of degrees as the angle-of-dip of the bedding, to give a new point.
5. Rotate the tracing paper back to its original position relative to the net (north on north) and read off the new direction by drawing a line from the central point, through the new pole position for the cross-bedding, to the circumference. This azimuth is the original orientation (direction-of-dip) of the cross-bedding, i.e. the direction to which the current was going.
6. The original angle-of-dip of the cross-bedding can be obtained by rotating the tracing paper so that the new pole position for the cross-bed is on the equatorial line, and then noting the number of degrees from the net centre to the new pole position.

See Figure 7.4 for a worked example.

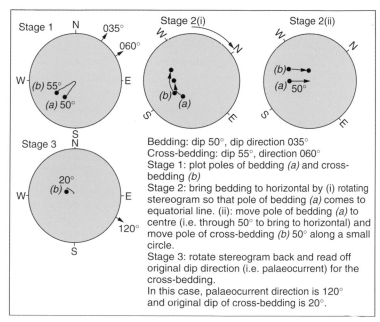

Bedding: dip 50°, dip direction 035°
Cross-bedding: dip 55°, direction 060°
Stage 1: plot poles of bedding (a) and cross-bedding (b)
Stage 2: bring bedding to horizontal by (i) rotating stereogram so that pole of bedding (a) comes to equatorial line. (ii): move pole of bedding (a) to centre (i.e. through 50° to bring to horizontal) and move pole of cross-bedding (b) 50° along a small circle.
Stage 3: rotate stereogram back and read off original dip direction (i.e. palaeocurrent) for the cross-bedding.
In this case, palaeocurrent direction is 120° and original dip of cross-bedding is 20°.

Figure 7.4 Correction of cross-bedding for tectonic tilt using stereographic projection. See text for more details.

Finally, a note of warning: if tilted strata are part of a fold whose axis is inclined (i.e. plunging), the fold axis must be brought to the horizontal, so changing the orientation of the tilt, before the tilt is removed. Having completed these corrections can you be certain that the palaeocurrent direction is that at its time of formation? Unfortunately not, for there may also have been rotations of the strata about a vertical axis; rarely can these be determined accurately.

7.3 Structures for Palaeocurrent Measurement

7.3.1 Cross-bedding

This is one of the best structures to use, but first determine what type of cross-bedding is present (see Section 5.3.3). If it has formed by the migration of subaqueous dunes and sand-waves (much is of this type) or aeolian dunes, then it is eminently suitable for palaeocurrent (or palaeowind) measurement. Check whether the cross-bedding is of the planar (tabular) or trough type (see Figure 5.16 and Section 5.3.2.2).

With *planar cross-bedding*, the palaeocurrent direction is simply given by the direction of maximum angle-of-dip. If the exposure is three-dimensional, or two-dimensional with a bedding-plane surface, there is no problem in measuring this directly. If there is only one vertical face showing the cross-bedding then taking readings is less satisfactory since it is just the orientation of the face that you are measuring, which is unlikely to be exactly in the palaeocurrent direction. Close scrutiny of the rock face may enable you to see a little of the cross-bed surface and so determine the actual dip direction. If even this is impossible there is no alternative but to measure the orientation of the rock surface.

With *trough cross-bedding*, it is essential to have a three-dimensional exposure or one with a bedding-plane section, so that the shape of the cross-strata is clearly visible and the direction down the trough axis can be measured accurately. Because of the shape of the trough cross-beds, vertical sections can show cross-bedding dipping up to 90° from the real current direction. Vertical sections of trough cross-bedding are thus unreliable for palaeocurrent measurements and should only be used as a last resort.

7.3.2 Ripples and cross-lamination

Palaeocurrent directions are easily taken from current ripples and cross-lamination. The asymmetry of the ripples (steeper leeside downstream) and the direction-of-dip of the cross-laminae are easily measured.

Ripples, however, and the cross-lamination they give rise to, are commonly produced by local flow directions, which do not reflect the regional palaeoslope. In turbidite beds, for example, cross-lamination within the bed may vary considerably in orientation and differ substantially from the palaeocurrent direction recorded by the sole structures. The cross-lamination forms when the turbidity current has slowed down and is wandering or meandering across the seafloor. In spite of their shortcomings, if there is no other more suitable directional structure present (cross-bedding or sole structures) it is always worth recording the orientation of the ripples and cross-lamination.

Wave-formed ripples are small-scale structures that record local shoreline trends and wind directions; their crest orientation should be measured, or if visible the direction-of-dip of internal cross-lamination.

7.3.3 Sole structures

Flute casts provide a current azimuth (see Section 5.2.1); some tool marks will also provide a current azimuth, if not at least a line of movement; groove casts provide a line of movement. Flute casts are generally all oriented in the same direction and so several measurements from one bed, together with measurements from other beds at the exposure, will be sufficient. With groove casts, there can be a substantial variation in orientation and a larger number (more than 20) should be measured, from which the vector mean can be calculated for each bed (see Section 7.4).

7.3.4 Preferred orientations of clasts and fossils, and imbrication

Pebbles and fossils with an elongation ratio of at least 3:1 can be aligned parallel-to or normal-to prevailing currents (see Figures 4.6, 6.7 and 6.8). Check for such a preferred orientation and if you either suspect or see its presence, measure and plot the elongation of a sufficient number of the objects (20 or more readings should do). In many cases you will obtain a bimodal distribution, with one mode, that parallel to the current, dominant. Pebbles, grains and fossils mostly give a line or trend of movement; with some fossils, for example orthoconic cephalopods, belemnites and high-spired gastropods, a direction of movement can be obtained (the pointed ends preferentially directed upstream). Stromatolite domes and columns may be asymmetric, with growth preferentially taking place on the side facing the oncoming currents and waves.

Flat pebbles in conglomerates and some fossils may show imbrication (see Section 4.4), whereby they are overlapping each other and dipping in an upstream direction (see Figure 4.7).

7.3.5 Other directional structures

Parting lineation records the trend of the current and presents no problem in its measurement. *Channels* and *scour structures* can also preserve the trend of the palaeocurrent. *Slump folds* record the direction of the palaeoslope down which slumping occurred (fold axes parallel to strike of palaeoslope, anticlinal overturning downslope). *Glacial striations* on bedrock show the direction of ice movement.

7.4 Presentation of Results and Calculation of Vector Means

Palaeocurrent measurements are grouped into classes of 10°, 15°, 20° or 30° intervals (depending on number of readings and variability) and then plotted on a rose diagram, choosing a suitable scale along the radius for the number of readings. For data from structures giving current azimuths, the rose diagram is conventionally constructed showing the current-to sense (in contrast to wind roses). For data from structures giving a trend, the rose diagram will be symmetrical. For presentation purposes, small rose diagrams can be placed alongside a graphic log at the location of the readings.

Although the dominant palaeocurrent (or palaeowind) direction will usually be obvious from a rose diagram, for accurate work it is necessary to calculate the mean palaeocurrent direction (i.e. the *vector mean*). It is also worth calculating the *dispersion* (or variance) of the data. Vector means and dispersions can only be calculated for a unimodal palaeocurrent pattern (Figure 7.1). However, if you have a bimodal distribution with two clear current directions, then you could separate the data into the two component parts and calculate the two vector means to find out the general direction of each current.

To deduce the vector means from structures giving azimuths, each observation is considered to have both direction and magnitude (the magnitude is generally considered unity but it can be weighted); the north-south and east-west components of each vector are then calculated by multiplying the magnitude by the cosine and sine of the azimuth respectively. Thus, make a table and simply look up (or use calculator) the cosine and sine of each palaeocurrent reading; add up each column and a division of the summed E-W components (sine values) by the N-S components (cosine values) gives the tangent of the resultant vector. Convert the tangent to an angle with the calculator; this is the vector mean – the average palaeocurrent direction.

$$\text{E-W component} = \sum n \sin \sigma$$

$$\text{N-S component} = \sum n \cos \sigma$$

$$\tan \overline{\sigma} = \frac{\sum n \sin \sigma}{\sum n \cos \sigma}$$

where:

σ = azimuth of each observation from $0°$ to $360°$.

n = observation vector magnitude, generally 1, but if data are grouped into classes ($0-15$, $16-30$, $31-45$, etc.), then it is the number of observations in each group.

$\overline{\sigma}$ = azimuth of resultant vector (i.e. vector mean).

If trends only can be measured then each observation, measured in the range $0-180°$, is doubled before the components are calculated:

$$\text{E-W component} = \sum n \sin 2\sigma$$

$$\text{N-S component} = \sum n \cos 2\sigma$$

$$\tan 2\overline{\sigma} = \frac{\sum n \sin 2\sigma}{\sum n \cos 2\sigma}$$

The magnitude (r) of the vector mean gives an indication of the *dispersion* of the data, comparable to the standard deviation or variance of linear data:

$$r = \sqrt{(\Sigma n \sin \sigma)^2 + (\Sigma n \cos \sigma)^2} \text{ for azimuthal data, and}$$

$$r = \sqrt{(\Sigma n \sin 2\sigma)^2 - (\Sigma n \cos 2\sigma)^2} \text{ for non-azimuthal data.}$$

To calculate the magnitude of the vector mean in terms of a percent (L):

$$L = \frac{r}{\Sigma n} \cdot 100$$

A *vector magnitude* of 100% means that all the observations have either the same azimuth or lie within the same azimuth group. In a vector magnitude of 0% the distribution is completely random. There would be no vector mean in this case.

For further discussion and methods for testing the significance of two-dimensional orientation distributions see Potter and Pettijohn (1977). Vector means can easily be calculated in the evenings after fieldwork and then entered on the geological map or graphic sedimentary log. The raw data should always be kept, however.

7.5 Interpretation of the Palaeocurrent Pattern

There are four types of palaeocurrent pattern (Figure 7.1):

- *unimodal* – where there is one dominant current direction;
- *bimodal bipolar* – two opposite directions;
- *bimodal oblique* – two current directions at an angle less than 180°;
- *polymodal* – where there are several dominant directions.

Analysis of the palaeocurrent pattern needs to be combined with a study of the lithofacies for maximum information. The features of the palaeocurrent (or palaeowind) pattern of the principal depositional environments – fluvial, deltaic, aeolian sand, shoreline-shallow shelf, and turbidite basin – are shown in Table 7.1.

In *fluvial facies*, palaeocurrents are best measured from the largest cross-beds to give the regional palaeoslope. Directions obtained from smaller structures, ripples and parting lineation, will generally show the minor currents of the river in low stage, and so not reflect the larger-scale palaeogeography. If lateral accretion surfaces are present, measure these too to deduce the direction of meandering. Braided-stream deposits tend to give palaeocurrent directions that have a lower dispersion; meandering-stream facies will show a wider range of directions (see Table 7.1).

Deltaic facies will give a variety of palaeocurrent patterns depending on the type of delta (lobate versus elongate) and the roles of fluvial and coastal processes. In a fluvial-dominated system, unimodal patterns will be obtained with dispersion depending on delta type. Where reworking of delta-front sands by marine processes is important, polymodal patterns may be obtained from wave and tidal effects.

Shoreline, shoreface and *shelf sandstones* are affected by wave, tidal and storm processes and so can have complex palaeocurrent patterns. Where tidal currents dominate, bipolar patterns may be obtained, but often one tidal current direction is stronger so that the pattern is asymmetrical. Tidal currents vary from parallel to oblique to normal to the shoreline. Storm waves and currents generate cross-beds usually directed offshore, but there may be a wide dispersion. Again, measure larger structures where possible (cross-beds, sole structures) rather than ripples, which may represent minor reworking.

In *basinal successions*, palaeocurrents from turbidites are often related to regional basin-ward directed palaeoslopes, but having

Table 7.1 *Palaeocurrent patterns of principal depositional environments, together with* best *and other directional structures.*

Environment	Directional structures	Typical dispersal patterns
Aeolian	Large-scale cross-bedding	Unimodal common, also bimodal and polymodal; dependent on wind directions/dune type
Fluvial	*Cross-bedding*, also parting lineation, ripples, cross-lamination, channels	Unimodal down palaeoslope, dispersion reflects river sinuosity
Deltaic	*Cross-bedding*, also parting lineation, ripples, channels	Unimodal directed offshore, but bimodal or polymodal if marine processes important
Marine shelf	*Cross-bedding*, also ripples, fossil orientations, flutes/grooves on bases of storm beds	Bimodal common through tidal current reversals but can be normal or parallel to shoreline; unimodal and polymodal patterns
Turbidite basin	*Flutes*, also grooves, parting lineation, ripples	Unimodal common, either downslope or along basin axis if turbidites; parallel-to-slope if contourites

reached the basin centre, deep-sea currents may then flow along the basin axis. It is useful to know the basin orientation and larger-scale tectonic context of the succession to interpret the palaeocurrent pattern. Flutes and grooves are the best structure to measure since a turbidity current begins to wander once it is slowing down so that structures within and upon the bed upper surface (cross-lamination and ripples) may not represent the main current direction. Slump folds are good for

5. Sedimentary Structures

6. Fossils in the Field

7. Palaeocurrent Analysis

8. Facies Identification

palaeoslope orientation. Some deep-water currents flow parallel to the contours of slopes, as in contourite deposits.

Desert sandstones can show very simple or highly complex palaeowind patterns as measured from large-scale cross-bedding. It depends on the nature of the sand deposit – large ridges (seif draas, see Figure 7.5) or sand seas (ergs) – the wind system and local topography. Some desert sandstones have unidirectional cross-beds from the effects of persistent trade winds. Palaeowind directions are not related to regional palaeoslope and interpretations of the palaeogeography need to be made with care.

Figure 7.5 *Cross-section through the top of a desert seif draa deposit which is one of more than 10 elongate sand-bodies extending for 10 kilometres, each 1–3 km wide. The overlying black bed is an organic–rich marl deposited when the sub-sea-level desert basin was flooded; it is overlain by shallow-water dolomite. Upper Permian, NE England.*

8

WHAT NEXT? FACIES IDENTIFICATION AND SEQUENCE ANALYSIS

8.1 Introduction

Having collected all the field data from the sedimentary succession it remains to interpret the information. Many studies of sedimentary rocks are concerned with elucidating the conditions, environments and processes of deposition. The field data are essential to such considerations. Other studies are concerned more with particular aspects of the rocks, such as the possibility of there being economic minerals and resources present, the origin of specific structures or the diagenetic history. Following the fieldwork, laboratory examination of the rocks is necessary in many instances, not least to deduce or confirm sediment composition and mineralogy.

8.2 Facies Analysis

If the aim of the study is to deduce the depositional processes and environments then, with all the field data at hand, the facies present within the succession should be identified. A *facies* is defined by a particular set of sediment attributes: a characteristic lithology, texture, suite of sedimentary structures, fossil content, colour, geometry, palaeocurrent pattern, and so on. A facies is produced by one or several processes operating in a depositional environment, although of course the appearance of the facies can be considerably modified by post-depositional, diagenetic processes. Within a sedimentary succession there may be many different facies present, but usually the number is not that great. Some facies may be repeated several or many times in a succession. A facies may also change vertically or laterally into another facies by a change in one or several of its characteristic features. In some cases

---

Sedimentary Rocks in the Field: A Practical Guide, Fourth Edition   Maurice E. Tucker
© 2011 John Wiley & Sons, Ltd

it will be possible to recognise *subfacies*, sediments that are similar to each other in many respects but that show some differences.

Facies are best referred to objectively in purely descriptive terms, using a few pertinent adjectives; examples could be cross-bedded, coarse sandstone facies or massive pebbly mudstone facies or nummulitic pack-grainstones. Facies can be numbered or referred to by letter (facies A, facies B, etc.), or a shorthand can be used that gives an indication of what the facies is like, that is a *lithofacies code* (see Section 2.6). In some circumstances, facies are referred to by their environment of deposition, such as braided stream facies or lagoonal facies, or by their depositional mechanism, such as turbidite sandstone facies or storm bed facies or microbialite facies. In the field and during the early stages of the study, facies should be referred to only in the descriptive sense. Interpretations in terms of process and/or environment can come later.

After logging and examining a succession in detail, look closely at the log with all the sediment attributes recorded and look for beds or units with similar features. Take the sedimentary structures first since these best reflect the depositional processes; then check the texture, lithology and fossil content. You will probably find that there are a relatively small number of distinct sediment types with similar attributes; these will be of the same facies. Name or number them for reference.

Once the various facies have been differentiated, make a table with their various features (name, code, typical thickness or thickness range, grain-size, sedimentary structures, fossils, colour, etc.). The facies can then be interpreted by reference to published accounts of modern sediments and ancient sedimentary facies. Many textbooks contain reviews of modern depositional environments, their sediments and their ancient analogues. See Recommended Reading for references.

Some facies are readily interpreted in terms of depositional environment and conditions, whereas others are not environmentally diagnostic and have to be taken in the context of adjacent facies. As an example, a fenestral peloidal lime mudstone will almost certainly have been deposited in a tidal-flat environment, whereas a cross-bedded coarse sandstone could be fluvial, lacustrine, deltaic, shallow-marine or even deep-marine, and could have been deposited by various processes. A number of depositional processes produce distinctive facies but can operate in several environments; for example, density currents depositing graded beds occur in lacustrine and marine basins, shallow and deep.

Facies interpretation is often facilitated by considering the *vertical facies succession*. Where there is a conformable vertical succession of facies, with no major breaks, the facies are the products of environments that were originally laterally adjacent. This concept has been appreciated since Johannes Walther expounded his Law of Facies in 1894. The vertical succession of facies is produced by the lateral migration of one environment over another (e.g. the progradation of a delta or tidal flat, the meandering of a stream). Where there are breaks in the succession, seen as sharp or erosional contacts between facies, then the facies succession need not reflect laterally adjacent environments but could well be the products of widely separated environments. Other environments whose sediments have been eroded could be represented by the break. There may have been some major change in the depositional conditions at the stratigraphic break, such as a rise or fall in relative sea level. See Section 5.3.1.1 on bedding planes.

In a sedimentary succession, you may well find that a group of facies occur together to form a *facies association*. The facies comprising an association were generally deposited in the same broad environment, in which there were several different depositional processes operating, distinct subenvironments or fluctuations in the depositional conditions. Think of a delta or submarine fan; there are several different processes operating there, depositing different sediment types (facies), but they are all related and so form a facies association.

8.3 Facies, Facies Models and Depositional Environments

From the study of modern depositional environments and their sediments, and their ancient equivalents, *facies models* have been constructed to represent and summarise the features of the depositional system and to show the lateral and vertical relationships between facies. These models facilitate interpretations of sedimentary rocks and permit predictions of facies distributions and geometries. However, it has to be remembered that facies models are just snapshots of an environment, and that depositional systems are dynamic; the facies model may only relate to a particular state of sea level, for example, or a particular climatic zone or latitude, or even a particular period in geological time. Very generalised facies models for the major depositional

5. Sedimentary Structures

6. Fossils in the Field

7. Palaeocurrent Analysis

8. Facies Identification

environments are presented later in this chapter, for the braided-stream (see Figure 8.15), meandering-stream (see Figure 8.16), deltaic (see Figure 8.18), siliciclastic shoreline (see Figure 8.20), deep-marine (see Figure 8.24), carbonate shelf (see Figure 8.25) and carbonate ramp environments (see Figure 8.26), to be considered in association with Tables 8.3–8.14, showing the features of the various facies.

From your facies data, it is possible for you to erect your own facies model. Make your interpretations of the depositional environment and subenvironments from the facies and think about their two- and three-dimensional arrangement. Make sketches, cross-sections and block diagrams to suggest the distribution of your facies and subfacies. Having done this, next think about the major controls on deposition: sea-level change, climate, tectonics, sediment supply/production and the biota (fossil record at the time). Next look for cycles and sequences in your succession of facies; see Section 8.4.

The principal features of the main sedimentary facies are given in Tables 8.3–8.14 at the end of this chapter. These are very generalised and are only intended to give a broad indication of the appearance of the various facies deposited in the common depositional environments. They are in no way complete since really it is not possible to summarise adequately the features of the various facies in a page or two, or in a simple table. Facies interpretation does require much more thought and industry than just looking at a few tables. Very detailed analyses of facies can now be made, and for help you should refer in the first instance to the textbooks cited in the Recommended Reading section and then to the scientific journals; also check the internet of course.

8.4 Cycle Stratigraphy and Sequence Stratigraphy

As noted elsewhere in this book, sedimentary rocks are commonly arranged into distinct units that are repeated several or many times in a succession. Thin repeated units, on a scale of 1–10 m, are usually referred to as sedimentary *cycles*, or in sequence stratigraphy terminology, *parasequences* (see Section 2.10). They were deposited over timespans of a few tens to hundreds of thousands of years. Parasequences are the building blocks of *sequences*, which are generally on the scale of many tens to hundreds of metres in thickness, deposited over a timespan of 0.5–3 million years (see Section 2.10). Table 8.1 gives the points to note when recording cyclic sediments.

Table 8.1 *Features to describe with metre-scale cyclic sediments and parasequences.*

(a) Cycle/parasequence boundaries

Look for evidence of exposure at top of cycle (e.g. palaeokarsts, potholes, palaeosoils, rootlets, coal, fenestrae, supratidal evaporites, vugs, collapse breccias)

If no exposure look for evidence of a break in sedimentation (e.g. intense bioturbation, hardground with encrusting fossils and borings)

Look for evidence of flooding at base of cycle (e.g. shale, phosphorite, glauconite, lag deposit, reworked pebbles and fossils)

(b) Internal structure of cycles/parasequences

Look for upward changes in lithofacies – e.g. limestone passing up into dolomite or gypsum, lime mudstone passing up into grainstone (or vice versa), mudrock passing up into sandstone

Look for upward changes in grain-size (coarsening-upward or fining-upward)

Look for upward changes in bed thickness (thickening-upward or thinning-upward)

(c) Stacking patterns of cycles/parasequences

Look at thickness of successive cycles; is each cycle thicker or thinner upwards?

Look at degree of exposure at tops of cycles: is it increasing or decreasing upwards through the stack of cycles?

Look at the facies within the parasequence stack: is there a long-term trend of shallowing or deepening? E.g. are the cycles more intertidal or subtidal up the stack?

Look for repeated packaging of the cycles into sets; e.g. are they organised into several/many packages of three to eight cycles?

Make a Fischer plot to show cycle thickness variation through the succession versus the average cycle thickness. Test for significance of plot (e.g. z-score)

8.4.1 Sedimentary cycle motifs

Metre-scale cycles vary considerably in composition and facies depending on the depositional environment. The majority of shallow-water, shoreline, shelf and platform cycles show a shallowing-upward trend, recorded by changes in lithologies, composition, grain-size, fossils and microfacies.

Some cycles are alternations of lithologies, for example mudrock-limestone in deeper-water basinal as well as shallower-water successions (Figure 8.1), limestone-gypsum or limestone-dolomite in carbonate platform interior successions, mudrock-sandstone in shallow- and deep-marine siliciclastic successions (Figures 8.2 and 8.3) and sandstone-mudrock in meandering-stream strata (see Figures 8.16 and 5.3). Mixed clastic-carbonate cycles also occur (Figure 8.4).

Other metre-scale cycles show systematic upward changes within the cycle, for example a coarsening-up of grain-size, as in deltaic (see Figures 8.2 and 8.18) and submarine fan-lobe mudrock to sandstone cycles (see Figures 8.3 and 8.24), or deeper-water lime mudstone to shallower-water grainstone cycles of carbonate ramps, as well as the opposite, a fining-up of grain-size as in meandering stream sandstone-to-mudrock cycles (see Figure 8.16), or high-energy shallow-water grainstone to tidal-flat fenestral lime mudstone of carbonate platform interiors (Figure 8.5).

Figure 8.1 *Metre-scale cycles of mudrock passing up into sandy limestone. Younging direction to the left. Cretaceous, Argentina.*

Figure 8.2 *Cycles consisting of mudrock overlain by sandstone. The top of the lowest sandstone has large-scale planar cross-bedding (to the right) from migration of a large sand-wave on a marine shelf. The top surface of this first sandstone is bioturbated and then sharply overlain by dark shale (a flooding surface). There is then a coarsening upwards to the next sandstone. This is the result of progradation of a small delta. A sharp surface on the second sandstone overlain by dark shale is also a flooding surface, and the mudrock passes up to a third sandstone, but this one has a sharp base as it is a channel sandstone of a delta distributary. Triangles indicate grain-size trends: inverted triangle = coarsening upwards. Height of section 15 m. Carboniferous, NE England.*

There may be systematic upward changes in the thickness of beds within cycles (thinning-up or thickening-up, as in Figure 8.3; see Section 8.4.3).

The presence of cycles within a succession can usually be seen in the field by close observation of the facies. In some cases the cycles will create steps or slope changes in the topography (Figure 8.6), as a

Figure 8.3 *Packaging of deep-water sandstones (composition is greywacke). Turbidite beds are arranged into coarsening-upward units 2–5 m in thickness consisting of some 5–10 individual event beds, each unit separated by 0.5–1 m of mudrock from the next. In some turbidite packages there is a suggestion of bed thickness increasing upwards. Younging direction to the right. Triangles indicate grain-size trends: inverted triangle = coarsening upwards. Cretaceous, California.*

result of some beds being eroded out more easily. In some cases, the cyclicity is only revealed after a detailed graphic log has been made or after some statistical tests have been applied (see textbooks cited in Recommended Reading). In some cases, it has to be said, the cyclicity is more apparent than real.

8.4.2 Cycle boundaries

Examine the boundaries between cycles carefully; distinct horizons may occur here. In many cases the top of a cycle is an exposure horizon; for example a palaeosoil (calcrete, seatearth or rootlet bed), palaeokarstic surface, desiccated microbial laminites or fenestral lime mudstones. It may be a sharp erosion surface. The tops of some cycles were not subaerially exposed, but show evidence of shallowing up to it, and a pause in deposition may be recorded in the form of intense bioturbation or a hardground with encrusting and boring organisms. See Figure 5.5 for the variety of bedding surfaces, which are often cycle boundaries.

The base of a cycle is usually a flooding surface. Thus there may be a thin conglomerate (a *basal lag*) of material reworked from the top of the underlying cycle with evidence of erosion (a sharp, scoured surface)

Figure 8.4 Mixed clastic-carbonate cycle: well-bedded transgressive shelf carbonates overlain by marine, then prodelta mudrocks, which pass up into deltaic sandstones, but here the coarsening-upward clastic package is cut into by a major fluvial channel, which shows lateral accretion (left to right) and fining upwards above. Height of cliff 35 m. Mid-Carboniferous, NE England.

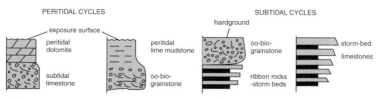

Figure 8.5 The spectrum of shallowing-upward carbonate units (cycles/parasequences); thicknesses are typically 0.75–2 m, but may reach more than 10 m.

Figure 8.6 Metre-scale carbonate platform interior cycles. Note the stepped hill-side profile. Cliff 300 m high. Jurassic, Empty Quarter, Yemen.

Figure 8.7 Top of one cycle in Figure 8.6 showing rubbly, pebbly, brecciated layer with blackened pebbles, all the result of subaerial exposure, succeeded by a sharp-based shallow subtidal, lagoonal limestone with reworked pebbles from below. Jurassic, Yemen.

beneath (Figure 8.7). There may be a thin muddy bed at the base of a cycle, reflecting the deeper-water conditions of the transgression initiating the new cycle. Across a flooding surface there is a deepening-up of facies.

8.4.3 Cycle stacking patterns and cyclostratigraphy

If a succession is cyclic, then there will probably be some systematic changes in the nature of the cycles up through the succession. It is important to document these as they will reflect the longer-term controls on deposition, primarily changes in accommodation space through time, and factors related to these. In many cases there are systematic variations in cycle thickness through a succession, cycles gradually increasing or decreasing in thickness upward (Figures 8.8 and 8.9). In some instances, cycles are arranged into packages of three to eight cycles, to form a *cycle set* (or *parasequence set*), with the thickness of each successive cycle decreasing or increasing upwards (e.g. Figures 8.9 and 8.10). In fact, there can be a whole hierarchy of cycles – cycle, cycle set, mesocycle set, megacycle set – in the succession.

Measure up the cycle thickness through a succession and look for any patterns (i.e. thinning-upwards or thickening-upwards). A *Fischer plot* is a useful way of presenting the data, especially for peritidal carbonate cycles, which shallow-up to sea level. In a Fischer plot, the

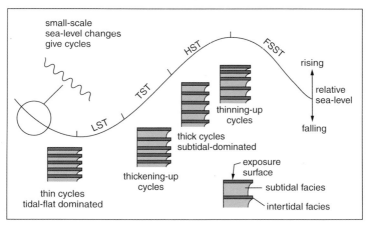

Figure 8.8 *The stacking patterns of parasequences within a sequence. Each individual metre-scale cycle is produced by one, high-frequency relative sea-level cycle, and the longer-term thickness pattern reflects the lower-frequency, longer-term change in accommodation space. The stacking patterns define the systems tracts of the sequence, as shown. LST, lowstand systems tract; TST, transgressive systems tract; HST, highstand systems tract; FSST, falling-stage systems tract.*

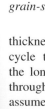

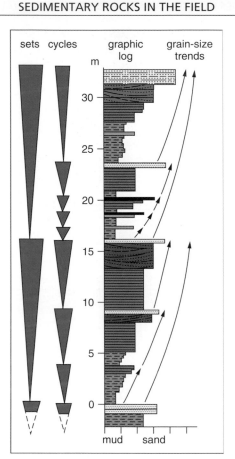

Figure 8.9 *An example of metre-scale cycles and their stacking pattern; a succession consists of coarsening-upward units (thin arrows show the cycles), and they group into packages (sets) based on the increasing grain-size and thickness.*

thickness of each successive cycle is plotted relative to the average cycle thickness (Figure 8.11). The pattern is interpreted to reflect the long-term change in accommodation space and relative sea level through time. Without more detailed study, however, it is dangerous to assume that each cycle represents the same length of time. Statistical

Figure 8.10 *Metre-scale parasequences (the beds) bundled into parasequence sets, shown by the arrows and highlighted by the lines of snow. Triassic, Brenta Dolomites, Italy.*

tests, calculation of z-scores, runs analysis and autocorrelation, can be applied to the thickness data of cycles to confirm whether the thickness patterns are random or not, and to help deduce the origin of the cyclicity (see Recommended Reading).

There may also be systematic changes in the internal structure of the cycles themselves and of the nature of the cycle boundaries, up through the succession. Look for variations in the facies of successive cycles (e.g. upward increasing/decreasing proportions of subtidal to intertidal facies; the incoming of sandy beds in a dominantly limestone succession). The degree of subaerial exposure at cycle tops may also increase or decrease up the succession. If the cycles are developed over a large area, then look for lateral changes in the facies of individual cycles if they can be correlated, or of the package of cycles as a whole if separate localities are involved.

The close study of metre-scale cycles, *cyclostratigraphy*, can help in the correlation of successions. In some cases, one specific cycle will show some particular features, such as a fossil band or well-developed exposure surface with a special colour, which will enable

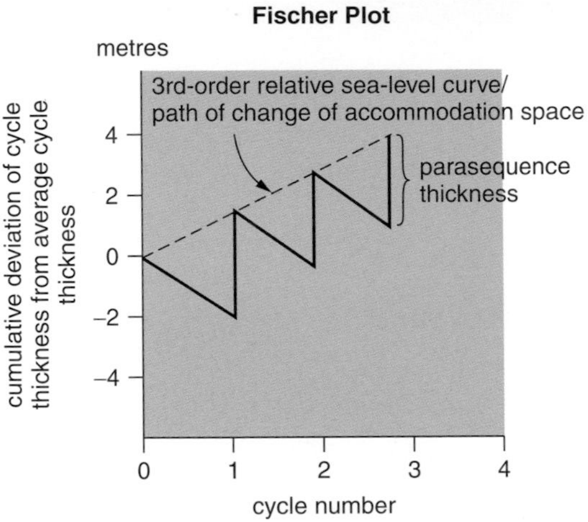

Figure 8.11 *A Fischer plot. From the total thickness of the succession calculate the average thickness of a cycle. Plot the average cycle thickness (in this example 2 m) as a diagonal line and then the thickness of each successive cycle as a vertical line. The horizontal scale is the cycle number through time. The vertical scale shows the deviation of cycle thickness through the succession from the average cycle thickness. Do not assume that each cycle was deposited over the same length of time. For cycles that shallow-up to sea level, the trend of the cycle thickness through time may reflect the longer term change in relative sea level and accommodation space. See appropriate books and papers for further information.*

correlations between outcrops. Once you get to know your strata and their characteristic features, you can then look for these in other out-crops; it is surprising how laterally extensive are bedding planes, facies, shell beds, and so on. In a more general way, the broad, upward changes in facies/thickness/grain-size, etc. through a cyclic succession can also be used for correlation.

The origin of the metre-scale cycles (parasequences) is of great interest, and useful information can be obtained by making the observations suggested here. *Autocyclic mechanisms* (sedimentary processes such as

Table 8.2 *Workflow model for sequence stratigraphy (see Catuneanu et al., 2010).*

Model-independent workflow

Make basic sedimentological-stratigraphical observations: facies, contacts, lapouts, stacking patterns, geometries, all as described in this book. Review biostratigraphic data for the succession

Delineation of key surfaces (unconformities, exposure horizons, flooding surfaces) and genetic units (forced regressive, normal regressive, transgressive)

Model-dependent choices

Choose the most appropriate surfaces for sequence boundaries

Choose the most appropriate sequence stratigraphic model

Name the key surfaces and systems tracts

Figure 8.12 *Large-scale shallowing-up succession, the highstand of a depositional sequence, from mud-dominated strata, with the gradual incoming of thicker limestone units, and then a 100 m-thick massive limestone at the top. Cretaceous, Ionian coast, Greece.*

tidal-flat progradation and lateral migration of meandering streams) give rise to cyclic sediments as do *allocyclic mechanisms* such as tectonic processes (e.g. jerky subsidence on extensional faults and in-plane stress changes) and sea-level change, such as through orbital forcing and solar insolation variations, and glacioeustasy (see Recommended Reading).

SEDIMENTARY ROCKS IN THE FIELD

Figure 8.13 *Major fluvial sandstone-filled channel (incised valley) cutting down into a coarsening-upward mud-to-sand deltaic succession, which occurs above a marine limestone. Inverted triangle indicates coarsening-upward unit; dashed line shows channel base. Mid-Carboniferous, NE England.*

The nature of sedimentary cycles does vary through the geological record, in response to major periods of tectonic activity and tectonic quiescence, and to the longer-term variations in global climate, namely periods of *icehouse* (late Precambrian/Cambrian, Permo-Carboniferous and late Tertiary-Quaternary) and *greenhouse* (mid-Palaeozoic and Triassic through early Tertiary). The amplitudes of rapid (tens of thousands of year) sea-level changes were much higher in the former state compared to the latter.

8.4.4 Cycle stacking and sequence stratigraphy

In terms of sequence stratigraphy, introduced in Section 2.10 with terms defined in Table 2.6, the *stacking pattern* of the cycles (parasequences) within a succession of metre-scale cycles determines the *systems tract* to which they belong as part of the *sequence*. For example, thickening-upward, subtidal-dominated parasequences capped by weak (or no) exposure horizons, reflecting increasing accommodation through time, typify the transgressive to highstand systems tract, whereas thinning-upward, inter/supra-tidal facies-dominated parasequences capped by well-developed emergence horizons, reflecting decreasing accommodation through time, are typical of late highstand, falling stage and lowstand systems tracts (see Figure 8.8).

Within many successions dominated by metre-scale cycles, there may not be one clear unconformity/cycle-top surface that can be taken as the sequence boundary, but there is a *sequence boundary zone* (SBZ) where

Table 8.3 *General features of fluvial facies.*

Deposition: is complex; alluvial systems include meandering streams with well-developed floodplains (see Figure 8.16), braided streams (see Figure 8.15) and alluvial fans. In the first, lateral migration of channels is characteristic, with muddy overbank sedimentation and sandy crevasse splays on floodplains. Channel processes dominate in braided streams, which may be gravel- or sand-dominated, and on alluvial fans, stream and sheet floods and debris flows occur.

Lithologies: from conglomerates through sandstones to mudrocks; thin intraformational conglomerates common; many sandstones are lithic or arkosic, compositionally immature to mature.

Textures: many stream-deposited conglomerates have a pebble-support fabric with imbrication; debris-flow conglomerates are matrix supported; most fluvial sandstones consist of angular to subrounded grains, with poor to moderate sorting, i.e. texturally immature to mature; some fluvial sandstones and mudrocks are red.

Structures: fluvial sandstones show tabular and trough cross-bedding, flat bedding + parting lineation, low-angle cross-beds (lateral accretion surfaces), channels and scoured surfaces; finer sandstones show ripples and cross-lamination; stream-deposited conglomerates are lenticular with flat bedding and crude cross-bedding; floodplain mudrocks are usually massive, possibly with rootlets and calcareous nodules (calcrete); they may contain thin, persistent sharp-based sandstones deposited from crevasse splays and floods.

Fossils: plants dominate (fragments or in situ), also fish bones and scales, freshwater molluscs, vertebrate tracks, some dwelling burrows.

Palaeocurrents: unidirectional, but dispersion depends on stream type: braided stream sandstones lower dispersion than meandering stream sandstones.

Geometry: sand bodies vary from ribbons to belts to fans.

(*continued overleaf*)

Table 8.3   *(continued)*

Facies successions and cycles: depend on type of alluvial
   system: alluvial fan strata may show an overall coarsening-up
   or fining-up depending on climatic/tectonic changes;
   meandering streams produce fining-upward cross-bedded
   sandstone units up to several metres thick with lateral
   accretion surfaces, interbedded with mudrocks, which may
   contain calcretes and thin persistent sandstones deposited by
   crevasse splays and floods (Figure 8.16); sandy braided
   streams produce lenticular and multi-storey cross-bedded
   sandstones with few mudrock interbeds (Figure 8.15).
Relevant sections: Sections 3.2, 3.3, 3.4, 5.2 and 5.3.
Relevant figures: Figures 4.7, 4.10, 4.12, 4.13, 5.3–5.18, 5.35,
   5.38, 5.61–5.63, 8.15 and 8.16.

Figure 8.14 *Large glacial dropstone of granite in well-bedded mudrock.*
Glaciomarine facies. Permian, Western Australia.

cycles tend to be thinner, more intertidal/subaerial facies are present, and
there is more evidence for exposure. In the same way, there may not be a
specific maximum flooding surface (mfs), but there will be a zone (maximum flooding zone, mfz) within the cyclic succession where cycles are
thicker and dominantly subtidal with little evidence of exposure.

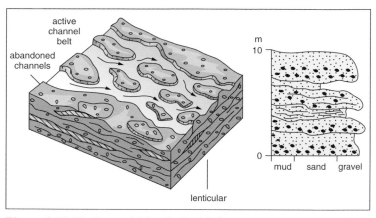

Figure 8.15 *Facies model for the braided-stream environment and typical succession consisting of lenticular sorted conglomerates and coarse sandstones.*

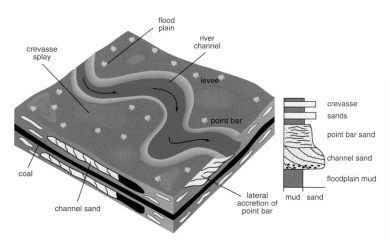

Figure 8.16 *Facies model for the meandering-stream environment and typical succession consisting of a fining-upward sandstone unit capped by mudrock produced by lateral migration of the stream. Such units vary in thickness from a few metres to a few tens of metres. Lateral accretion surfaces may occur within the sandstone member.*

Table 8.4   *General features of aeolian facies.*

Deposition: wind-blown sand is typical of deserts but also occurs along marine shorelines.

Lithology: clean (matrix free) quartz-rich sandstones, no mica. Carbonate dunes also (aeolianites).

Textures: well-sorted, well-rounded sand grains ('millet-seed'); possibly with frosted (dull) appearance; sandstones commonly stained red through hematite; pebbles may be wind-faceted.

Structures: dominantly large-scale cross-bedding (set heights several to several tens of metres); cross-bed dips up to 35°; reactivation surfaces and master bounding surfaces; grain-fall and grain-flow laminae; pin-stripe lamination.

Fossils: rare, possibly vertebrate footprints and bones, and plant rootlets.

Sand-body geometry: laterally extensive sheets if sand seas (ergs), and more elongate ridge-like geometries if seif draas.

Facies associations: water-lain sandstones and conglomerates (fluvial/flash-flood) may be associated; also playa-lake mudrocks and evaporites and arid-zone soils (calcretes).

Relevant sections: Sections 3.2.1, 4.5, 5.3.2.3 and 5.3.3.11.

Relevant figures: Figures 3.4, 5.14 and 5.32–5.34.

Table 8.1 gives the main features to look for when describing parasequences and cyclic sediments.

8.4.5   Sequence stratigraphy in the field

It is very popular these days to divide a succession into sequences (see Section 2.10 and Recommended Reading suggestions and Figure 1.1 for an example), although there are several models in the literature and much terminology. However, to make a sequence stratigraphic interpretation a lot of biostratigraphical and sedimentological information is required from the sedimentary basin as a whole. Observations from a single locality can often be thought of in terms of sequence stratigraphy but all ideas would need to be confirmed and likely modified after the examination of other localities in the region and a thorough understanding of the basin's stratigraphy and evolution.

Table 8.5 *General features of lacustrine facies.*

Deposition: in lakes, which vary in size, shape, salinity and depth. Waves and storm currents important in shallow water; turbidity currents and river underflows in deeper water. Biochemical and chemical precipitation common. Strong climatic control on lake sedimentation.

Lake types: permanent, perennial and ephemeral; hypersaline and freshwater; stratified and non-stratified; bench and ramp margins; carbonate, evaporitic and siliciclastic lakes.

Lithologies: diverse including conglomerates through sandstones to mudrocks, limestones (oolitic, micritic, bioclastic, microbial), marls, evaporites, cherts, oil shales and coals.

Structures: wave-formed ripples, desiccation cracks, rainspots and stromatolites common in lake shoreline sediments; spring deposits of calcareous tufa and travertine; rhythmic lamination, possibly with syneresis cracks, typical of deeper-water lake deposits, together with interbedded graded sandstones of turbidity current origin.

Fossils: non-marine invertebrates (especially bivalves and gastropods); vertebrates (footprints and bones); plants, especially algae.

Facies successions and cycles: reflect changes in water level through climatic or tectonic events; shallowing-upward cycles common, capped by exposure and/or pedogenic horizons.

Facies associations: fluvial and aeolian sediments commonly associated; soil horizons may occur within lacustrine sequences; mottled and marmorised palustrine mudrocks and limestones.

Relevant sections: Sections 5.3 and 5.5.6.2.

Relevant figures: Figures 5.8, 5.9–5.12, 5.19–5.22, 5.39–5.40, 5.49–5.51 and 5.61–5.63.

Sequence stratigraphy divides a succession on the basis of *key surfaces* into systems tracts (see Section 2.10.2, Figure 2.6 and Table 2.6). The sediments of the various systems tracts usually show particular facies successions – e.g. deepening-up in the transgressive systems tract (TST), shallowing-up in the highstand systems tract (HST) – or if the

5. Sedimentary Structures

6. Fossils in the Field

7. Palaeocurrent Analysis

8. Facies Identification

Table 8.6   *General features of pedogenic facies.*

Deposition: pedogenesis can take place at emergence horizons, on abandonment surfaces, at unconformities and cycle and sequence boundaries.

Lithologies: limestone (calcrete/caliche), dolomite (dolocrete), sandstone (seatearth, ganister, silcrete), mudrock (fireclay, seatearth), karstic breccias.

Texture: fine-grained mostly, also pisolitic, peloidal, mottled, marmorised.

Structures: massive, laminated, nodular, intraclastic, rhizocretions, sheet cracks, tepees, palaeokarstic surfaces, potholes/vugs.

Fossils: plants common, especially rootlets; others rare: non-marine vertebrates and invertebrates.

Geometry: generally sheet-like.

Facies sequences: pedogenic facies typically occur at the top of cycles, e.g. meandering-stream fining-upward cycles, deltaic and shoreline coarsening-upward cycles, and carbonate shallowing-upward cycles, and within lacustrine strata, generating the palustrine facies.

Relevant sections: Sections 3.2.1, 4.8, 5.5.6.2 and 5.6.4.

Relevant figures: Figures 5.61–5.63.

succession is composed of small-scale cycles (parasequences), then the systems tracts are defined on the stacking patterns and facies of the parasequences (see Section 8.4.4) and the recognition of regressive and transgressive trends. Sequence stratigraphy has become quite complicated in recent years with an increase in terminology and a range of models available. However, there are basic features that can be recognised and described objectively in the field (as well as from seismic, well logs and core), should be free from any bias, and from which the appropriate sequence stratigraphic model can be applied; the systems tract bounding surfaces can then be labelled accordingly and sequences themselves defined. Thus the workflow (Table 8.2) is to document the model-independent features of a succession first, and then to apply the most appropriate model.

Table 8.7   *General features of ancient (pre-Quaternary) glacial facies.*

Deposition: takes place in a wide variety of environments: beneath glaciers of various types, in glacial lakes, on glacial outwash plains and glaciomarine shelves and basins, and by a variety of processes including moving and melting glaciers, meltwater streams, meltwater density currents, debris flows and icebergs. See Table 8.8 for Quaternary till types.

Continental glacial environments: grounded-ice, glaciofluvial, glaciolacustrine – proglacial and periglacial lakes, cold-climate periglacial facies.

Glacial marine environments: proximal/shoreline, shelf, deep-water facies.

Lithologies: variety of conglomerates – polymictic muddy to pebbly conglomerates (diamictites and mixtites, which may be tillites), sandstones, muddy sediments with dispersed clasts (dropstones).

Texture: poorly sorted, matrix-supported conglomerates (diamictites) to better sorted clast-supported conglomerates where reworking/resedimentation has taken place; angular clasts possibly with striations and facets, and elongate clasts possibly showing preferred orientation.

Structures: diamictites/tillites generally massive but some layering may occur; rhythmically laminated ('varved') muddy sediments common (possibly with dropstones); fluvioglacial sandstones show cross-bedding, cross-lamination, flat-bedding, scours and channels. Striated pavements beneath continental tillites.

Fossils: generally absent (or derived), except in glaciomarine sediments.

Geometry: tillites lenticular to laterally extensive.

Facies sequences and associations: usually no repeated sequences but apparently random succession of tillites, fluvioglacial and more glaciolacustrine sediments; however, alternating periods of glaciation (tillites) and deglaciation (shallow-marine sandstones) may give cycles; debrites and turbidites associated with glaciomarine facies.

Relevant section: Section 4.6.

Relevant figures: Figures 4.9, 4.11 and 8.14.

5. Sedimentary Structures

6. Fossils in the Field

7. Palaeocurrent Analysis

8. Facies Identification

Table 8.8 *Main types of glacial till, as recognised in Quaternary successions, and the features of their deposits.*

Till type	Clast fabric	Clast shape	Deposit	Structures
Supraglacial till	Weak, chaotic	Generally angular and fresh	Diamicton with stratified units, remobilised to debrites	Deformation structures common, faults, slumps
Subglacial till: lodgement till	Strong fabric in ice-flow direction	Generally rounded and abraded	Usually homogeneous diamicton, sharp base, extensive	Structureless, joints, shears
Subglacial till: deforming bed till	Medium	Generally rounded	Homogeneous diamicton	Folds
Subglacial till: melt-out till	Weak to strong	Generally rounded	Homogeneous and stratified diamicton	Structureless, some bedding, may be deformed
Flow till: the above tills affected by gravity-flows	Weak to strong	Any shape	Stratified diamicton	Massive to graded, bedding, folds

FACIES IDENTIFICATION AND SEQUENCE ANALYSIS

Table 8.9 *General features of deltaic facies.*

Deposition: complex; there are several types of delta (lobate and elongate especially), and many deltaic subenvironments (distributary channels and levees, swamps and lakes, mouth and distal bars, interdistributary bays and prodelta slope). Many ancient deltas were river dominated but reworking and redistribution of sediment by marine processes was important.

Lithologies: mainly sandstones (compositionally immature to mature, commonly lithic) through muddy sandstones, sandy mudrocks to mudrocks; also coal seams and sideritic ironstones.

Textures: not diagnostic (texturally immature to mature); average sorting and rounding of sand grains.

Structures: cross-bedding of various types in the sandstones, flat-bedding and channels common. Finer sediments show flaser and wavy bedding and are heterolithic. Some sediments contain rootlets (seatearths, ganisters); nodules of siderite. Bioturbation and trace fossils common.

Fossils: marine fossils in some mudrocks and sandstones, others with non-marine fossils, especially bivalves. Plants common.

Palaeocurrents: mainly directed offshore but may be shore-parallel or onshore if much marine reworking with waves and tides.

Geometry: sand bodies vary from ribbons to sheets depending on delta type.

Facies successions and cycles: typically consist of coarsening-upward unit (mudrock to sandstone), through delta progradation, capped by a seatearth and coal (Figure 8.18); there are many variations, however, particularly at the top of such units.

Relevant sections: Sections 3.2, 3.4, 5.2 and 5.3.

Relevant figures: Figures 5.3–5.23, 5.35, 5.58, 8.2, 8.4, 8.13 and 8.18.

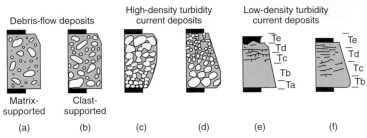

Debris-flow deposits   High-density turbidity current deposits   Low-density turbidity current deposits

Matrix-supported   Clast-supported

(a)   (b)   (c)   (d)   (e)   (f)

Figure 8.17 The range of deep-water sediment gravity-flow deposits: debrites and turbidites. These beds range from a few centimetres to a metre or more in thickness. Ta, Tb, Tc, Td and Te refer to the 'Bouma' divisions of a classic turbidite.

Three basic types of stratigraphic facies pattern (genetic units) are recognised for a sequence stratigraphic analysis, defined by specific stratal stacking patterns and separated by their bounding surfaces:

1. a *normal regressive unit* – progradation with aggradation driven by sediment supply, generating prograding, offlapping strata, showing shallowing up; typical of highstand and lowstand deposition;
2. a *forced regressive unit* – progradation driven by base-level fall; a down-stepping, prograding, offlapping package; typical of late highstand to falling stage to early lowstand systems tracts;
3. a *transgressive unit* – retrogradation, backstepping, driven by base-level rise; onlapping, deepening-upward unit, typical of transgressive systems tract.

Examining the facies patterns and looking for these larger-scale vertical and lateral changes and relationships is necessary to assign systems tracts. Figure 8.12 shows a large-scale shallowing-up carbonate succession, with increasing thickness of carbonate units up to a massive thick (~100 m) unit at the top. This whole succession would be a highstand systems tract fomed through major progradation.

The *key surfaces* have distinctive sedimentological features, which can be observed in the field, and these are the bounding surfaces to the systems tracts. These are noted here, but again, remember that it

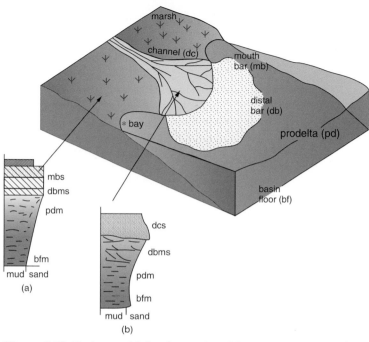

Figure 8.18 *Facies model for the marine deltaic environment and two typical successions: (a) coarsening-upward (c.u.) unit produced by delta progradation during a sea-level stillstand and capped by coal; and (b) a c.u. unit cut by a distributary channel sandstone. Thicknesses range from 10 to 30 m or more. There can be much variation in these units, particularly towards the top if interdistributary bays are developed. Suffixes: m, mud; ms, muddy sand; s, sand.*

is necessary to consider the succession on a larger scale than an outcrop, taking into account the regional sedimentology, biostratigraphy and the relationships between sedimentary units (onlap, downlap, etc.; see Section 5.7) to be sure that your interpretations are realistic.

In sequence stratigraphy, the key surfaces are the *subaerial unconformity-correlative conformity* (which is the *sequence boundary* in the 'classic' model, Figure 2.6), *transgressive surface* (ts) (base

Table 8.10 *General features of shallow-marine siliciclastic facies.*

Deposition: takes place in a variety of environments and subenvironments including tidal flat, beach, barrier island, lagoon, shoreface and nearshore to offshore shelf. Waves, tidal and storm currents are the most important processes.

Lithologies: sandstones (compositionally mature to supermature, may be quartz arenites) through muddy sandstone, sandy mudrocks to mudrocks; also thin conglomerates. Glauconite in greensands.

Textures: not diagnostic although sandstones generally texturally mature to supermature.

Structures: in the sandstones: cross-bedding, possibly herring-bone, reactivation surfaces, flat-bedding (in truncated low-angle sets if beach), wave-formed and current ripples and cross-lamination, flaser and lenticular bedding, bundled cross-beds if tidal, mud drapes; HCS and SCS if storm waves, thin graded sandstones of storm-current origin; mudrocks may contain pyrite nodules; bioturbation and various trace fossils common – the latter reflecting local energy level and depth.

Fossils: marine faunas with diversity dependent on salinity, level of turbulence, substrate, and so on.

Palaeocurrents: variable, parallel-to and normal-to shoreline, unimodal, bimodal or polymodal.

Geometry: linear sand bodies if barrier or beach, sheet sands if extensive epeiric-sea platform.

Facies successions and cycles: vary considerably depending on precise environment and sea-level history (rising or falling); coarsening-up, shallowing-up units from shoreline progradation (Figure 8.20).

Facies associations: limestones, ironstones and phosphates may occur within shallow-marine siliciclastic facies.

Relevant sections: Sections 3.2, 3.4, 5.3 and 5.6.

Relevant figures: Figures 5.4–5.32, 5.68–5.79 and 8.20.

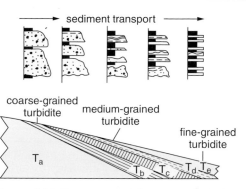

Figure 8.19 *Downcurrent changes in turbidites.*

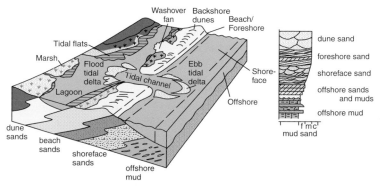

Figure 8.20 *Facies model for marine shoreline environments and the typical coarsening-upward succession developed through progradation of a beach/barrier shoreline during a sea-level stillstand. Typical thickness 10 m or more. Abbreviations: f, m and c = fine, medium and coarse (sand), respectively.*

of the TST) and the *mfs* (separating the TST from the HST/RST (regressive systems tract). Also important is the *condensed section* (CS), the more distal (basinward) equivalent of the mfs, which normally includes some of the upper TST and lower HST (see Section 2.10).

8.4.5.1 Subaerial unconformity-correlative conformity

The subaerial part is usually a distinctive surface that shows evidence of significant erosion and extended exposure (e.g. a palaeokarst or palaeosoil). There is usually a biostratigraphic gap, and underlying strata may be truncated by the unconformity. There will usually be a significant and abrupt facies change across the boundary; in many cases

Table 8.11 *General features of deeper-marine siliciclastic facies.*

Deposition: takes place on submarine slopes, submarine fans and aprons, in basins of many types, particularly by turbidity currents, debris flows, contour currents and deposition from suspension.

Lithologies: sandstones (generally compositionally immature to mature, commonly greywacke in composition) and mudrocks (hemipelagic); also conglomerates (pebbly mudstones).

Texture: not diagnostic; sandstones may be matrix-rich; conglomerates mostly matrix-supported and of debris-flow origin.

Structures: in turbidite sandstones: graded bedding (interbedded with hemipelagic mudrocks), may show 'Bouma' sequence of structures (Figures 8.17, 8.19 and 8.22), sole marks common, 5–100 cm thick; some sandstones may be massive. Contourites: muddy and sandy siltstones, gradational upper and lower contacts, reversely graded lower parts, normally graded upper parts, bioturbated, some cross-lamination, 10–30 cm thick. Hemipelagic mudrocks may be finely laminated or bioturbated. Channels, perhaps on large-scale, also slump, slide and dewatering structures.

Fossils: mudrocks chiefly contain pelagic fossils; interbedded sandstones may contain derived shallow-water fossils.

Palaeocurrents: in turbidite sandstones – variable, may be downslope or along basin axis, best measured on sole marks.

Facies successions and cycles: turbidite successions may show upward-coarsening and upward-thickening of sandstone beds (Figure 8.3), or upward fining and thinning.

Relevant sections: Sections 3.2, 5.2 and 5.3.

Relevant figures: Figures 5.1, 5.2, 5.7, 5.52, 5.6, 8.3, 8.17, 8.19 and 8.21–8.24.

Figure 8.21 *Turbidites (well-bedded, thinner beds) and a thick bouldery debrite, 3 m thick. Cambrian, Newfoundland, Canada.*

Figure 8.22 *A turbidite bed showing ABC divisions: a lower coarse massive division (A), overlain by a finer-grained parallel-laminated division (B) and then the upper part consisting of cross-laminae with a convolution at the top (C division). This turbidite is a bioclastic limestone, 25 cm thick. Devonian, SW England.*

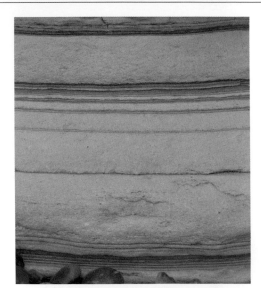

Figure 8.23 *Carbonate turbidite beds showing upward decrease in thickness. Height of section 30 cm. Upper Permian, NE England.*

it will be a change from marine facies (HST) below to non-marine facies [falling stage systems tract (FSST)/lowstand systems tract (LST)] above the boundary, or in more basinal locations where the correlative conformity occurs, from deeper-water, finer-grained facies of the highstand (below) to shallower-water, coarser-grained facies of the falling stage/lowstand (above), both indicating a downward (basinward) facies shift as a result of the negative accommodation that created the unconformity. In more proximal (landward) areas, however, deeper-water facies of the TST may well overlie the unconformity if, as is often the case, no falling stage/lowstand sediments were deposited there. Basinwards the unconformity will pass into a conformity that will show little or no evidence of erosion. Subaerial unconformities-correlative conformities will have regional, basinwide extent, and may possibly be correlatable with unconformities in other basins.

Associated with subaerial unconformities are commonly *incised valley fills* (IVFs). These, usually linear, major valleys are cut during the time of decreasing accommodation (forced regression/falling stage) and

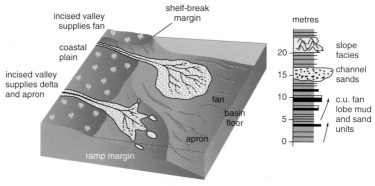

Figure 8.24 *Facies models for the deep-marine turbidite fan and apron environments developed at a shelf-break and ramp margin respectively, and the typical coarsening-upward succession developed. Such environments are best developed during the falling stage and lowstands of sea level.*

filled during the lowstand and subsequent TST. IVFs have sharp basal surfaces cutting down several to tens of metres; coarse fluvial clastics will fill the lower part, passing up into finer estuarine and marine sandstones. Major soil horizons are often laterally equivalent to IVFs, formed on the interfluves. An example is shown in Figure 8.13.

A another type of unconformity recognised in carbonate strata is the *drowning unconformity*, which is again a prominent horizon but one where deep-water facies overlies shallow-water limestone. The surface of the carbonate platform may have been exposed first, before the drowning, so that a palaeokarst is present. Drowning surfaces may be mineralised or encrusted with phosphate.

8.4.5.2 Transgressive surface (ts) (also called maximum regressive surface and transgressive ravinement surface)

This surface will usually show a major facies change from shallower-water or subaerial facies below to deeper-water facies above (which may be widespread), reflecting the increased water depth and relative rise in sea-level/positive accommodation. The ts is the first significant marine flooding surface across a shelf. A *lag deposit* (a pebble bed, bone bed, concentration of fossils) may occur upon the surface, which is usually

Table 8.12 *General features of shallow-marine carbonate facies.*

Deposition: takes place in a variety of environments and subenvironments: tidal flats, beaches, barriers, lagoons, nearshore to offshore shelves and platforms, epeiric shelf seas, submarine sand shoals and reefs (shelf-margin and patch reefs especially). Biological and biochemical processes largely responsible for formation and deposition of sediment, although physical processes of waves, tidal and storm currents important. Carbonate shelves, with steep shelf margin dominated by reefs and sand bodies (Figure 8.25), distinct from carbonate ramps with gentle offshore gradient, dominated by shoreline sands and offshore muds with storm beds (Figure 8.26).

Lithologies: many types of limestone, especially oolitic and skeletal grainstones, skeletal packstones-wackestones, mudstones and boundstones; also dolomites. Limestones may be silicified. Evaporites, especially sulphates (or their replacements) may be associated.

Textures: diverse.

Structures: great variety, including cross-bedding, ripples, desiccation cracks, stromatolites and microbial laminites, fenestrae, stromatactis and stylolites; reef limestones: massive and unbedded, many organisms in growth position.

Fossils: vary from diverse and abundant in normal-marine facies to restricted and rare in hypersaline or hyposaline facies.

Palaeocurrents: variable: parallel and normal to shoreline.

Facies successions and cycles: many types but metre-scale shallowing-upward cycles common in platform successions (Figure 8.6).

Relevant sections: Sections 3.5, 5.3, 5.4, 5.6 and 6.1–6.4.

Relevant figures: Figures 3.7–3.18, 5.41–5.51, 5.61–5.66, 5.69–5.79, 5.83–5.85, 6.1–6.10, 8.1, 8.4–8.12, 8.21–8.23, 8.25 and 8.26.

CARBONATE RIMMED SHELF			slope	basin	
subaerial	protected		maximum wave action	below fairweather wave-base	
supratidal carbonates	lagoonal and tidal flat carbonates		reefs and carbonate sand bodies	resedimented carbonates	shales/ pelagic limestones
mudstones	wackestones–mudstones		boundstones/ grainstones	grain/rud/float/ wackestones	mudstones

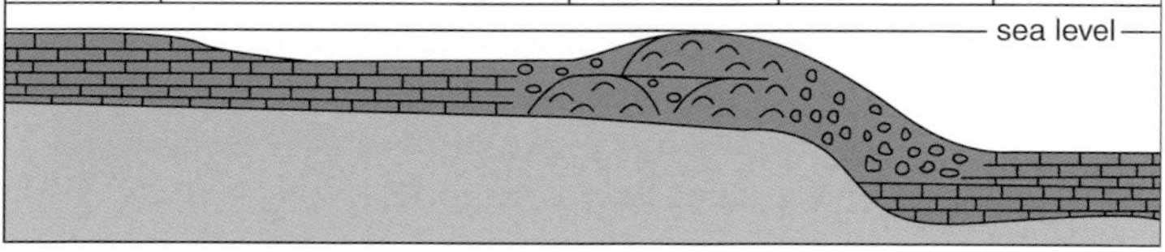

Figure 8.25 *Facies model for the carbonate shelf environment.*

CARBONATE RAMP				
inner ramp			mid ramp	outer ramp
protected/subaerial		wave-dominated	below fairweather wave-base	below storm wave-base
lagoonal–tidal flat–supratidal carbonates, sabkha evaporites, palaeosoils, palaeokarsts		beach-barrier/ strandplain/sand shoals, patch reefs	thin-bedded limestones, storm deposits ± mud mounds	shales/ pelagic limestones
wackestones–mudstones		grainstones	grain/rud/float/ wackestones	mudstones

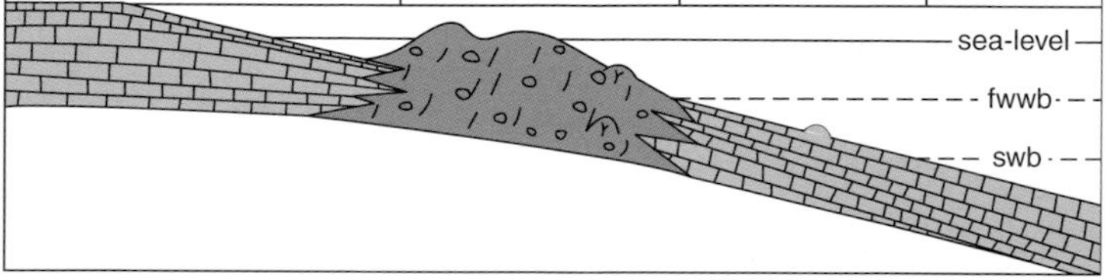

Figure 8.26 *Facies model for the carbonate ramp environment. fwwb, fairweather wave-base; swb, storm wave-base.*

Table 8.13 *General features of deeper-water carbonates and other pelagic facies.*

Deposition: takes place in deeper-water epeiric seas, outer shelves and platforms, submarine slopes, in basins of many types and on ridges and banks within basinal areas. Deposition from suspension and by resedimentation processes.

Lithologies: pelagic limestones are usually fine-grained with a dominantly pelagic fauna; limestone turbidites are fine-to-coarse grained and consist largely of shallow-water fossils or ooids; cherts, phosphorites, iron-manganese nodules, hemipelagic mudrocks associated.

Structures: pelagic limestones: commonly nodular, hardgrounds common together with sheet cracks and neptunian dykes, stylolites common; limestone turbidites: graded bedding and other structures (sole and internal) as in Figure 8.17 although usually less well developed; bedded cherts: may be graded and laminated. Pelagic sediments may be slump-folded and brecciated.

Fossils: pelagic fossils dominate; derived shallow-water fossils in limestone turbidites.

Facies successions and cycles: no typical sequences; pelagic facies may overlie or underlie turbidite successions or follow platform carbonates. Pelagic facies may be associated with volcaniclastic sediments and pillow lavas. Common are sub-metre-scale rhythms of alternating clay-poor/clay-rich limestone, or limestone-mudrock, and so on.

Relevant sections: Sections 3.5 and 3.8.

Relevant figures: Figures 3.20–3.22, 5.8, 5.77, 5.78, 6.7, 6.10, 6.11, 8.3, 8.21 and 8.22.

sharp, as a result of the erosion (ravinement) during the transgression. New fossil species may occur, and overall the sediments above the ts will generally deepen upwards.

8.4.5.3 Maximum flooding surface (mfs)

This surface should occur within the deepest-water or most marine facies of the sequence, since it represents the time of maximum

Table 8.14 *General features of volcaniclastic facies (see Section 3.11).*

Deposition: takes place in subaerial and submarine (shallow or deep) environments by pyroclastic fall-out, volcaniclastic flows such as ignimbrites and lahars, and the tephra may be reworked and resedimented by waves, tidal, storm and turbidity currents.

Lithologies: tuffs, lapillistones, agglomerates and breccias.

Textures: diverse, include welding in ignimbrites and matrix-support fabric in lahar deposits.

Structures: include grading in air-fall tuffs, current and wave structures in reworked and redeposited tuffs (epiclastics), planar and cross-bedding (including antidunes) in pyroclastic flow tuffs.

Fossils: do occur although rarely.

Facies successions: well-developed eruptive units may show air-fall tuffs overlain by pyroclastic flow deposits, capped by fine air-fall tuff.

Facies associations: submarine volcaniclastics commonly associated with pillow lavas, cherts and other pelagic sediments.

Relevant sections: Sections 3.11 and 5.3.3.15.

Relevant figures: Figures 3.23–3.30.

5. Sedimentary Structures

6. Fossils in the Field

7. Palaeocurrent Analysis

8. Facies Identification

transgression. It is generally the most significant flooding surface of the succession. It may not be an actual bedding plane but occur within, say, an organic-rich mudrock a few metres thick, with shallower-water facies below (of the TST) and above (of the HST). There may be intense bioturbation at the mfs reflecting the low rate of sedimentation, and glauconite and phosphate may occur there too.

In more distal parts of the basin, the upper part of the TST, the mfs and the lower part of the HST may be represented by a *condensed section*. This is a sediment-starved part of the succession and so it may be: an intensely bioturbated bed(s), a hardground from seafloor cementation, or an organic-rich mudrock; sediments may be impregnated with minerals such as glauconite, phosphorite, berthierine/chamosite or pyrite; there may be a distinctive colour and the unit may be correlatable over a wide area. A condensed section may be enriched in fossils and microfossils with high abundance and diversity, and there may be a stratigraphic gap (hiatus).

The assignment of facies to particular *systems tracts* depends on recognising the bounding surfaces noted above and on the nature of the facies themselves and their vertical and lateral development between the key surfaces. If the succession consists of metre-scale cycles, the key surfaces may not be individual horizons, but they may occur within a package of cycles (parasequences) where there is a change in their character and stacking pattern (see Section 8.4.4).

To reiterate, the sequence stratigraphic interpretation of a succession requires much detailed fieldwork over the region and thoughtful analysis of the data, as well as attention to the biostratigraphy and correlation. See Recommended Reading.

Figure 8.27 *Megabreccia deposit consisting of metre-scale blocks of shallow-water limestone emplaced into basinal mudrocks through platform-margin collapse. Cretaceous, Aravis, France.*

RECOMMENDED READING

Books Containing Sections on Relevant Techniques

Barnes, J.W. and Lisle, R.J. (2003) *Basic Geological Mapping*. John Wiley & Sons Ltd, Chichester, 196 pp.

Bhattacharyya, A. (2000) *Analysis of Sedimentary Successions: A Field Manual*. AA Balkema, Rotterdam.

Blackbourne, G.A. (2000) *Cores and Core Logging for Geologists*. Whittles Publishers, Caithness, 113 pp.

Bouma, A.H. (1969) *Methods for the Study of Sedimentary Structures*. Wiley-Interscience, New York, 458 pp.

BSI (British Standards Institute) (1981) *Code of Practice for Site Investigations*. BS 5930, 140 pp.

Coe, A.L. (Editor) (2010) *Geological Field Techniques*. Wiley, Chichester, 323 pp.

Goldring, R. (1991) *Fossils in the Field*. Longman, Essex, 218 pp.

Jones, A.P., Tucker, M.E. and Hart, J.K. (eds) (1999) *The Description and Analysis of Quaternary Stratigraphic Field Sections*. Technical Guide 7, Quaternary Research Association, 293 pp.

Tucker, M.E. (ed.) (1988) *Techniques in Sedimentology*. Blackwells, Oxford, 408 pp. See Chapter 2 by John Graham, 5–62 pp, on field techniques.

General Textbooks Concerned with Depositional Environments, Facies Analysis and Sedimentary Basins

Allen, P.A (1997) *Earth Surface Processes*. Blackwell Science, Oxford, 404 pp.

Allen, P.A. and Allen, J.R. (2005) *Basin Analysis: Principles and Applications*. Blackwell Publishing, Oxford, 549 pp.

Boggs, S. (2009) *Petrology of Sedimentary Rocks*. Cambridge University Press, 660 pp.

Einsele, G. (2000) *Sedimentary Basins*. Springer-Verlag, Berlin, 628 pp.

Flügel, E. (2004) *Microfacies of Carbonate Rocks*. Springer-Verlag, Berlin.

Leeder, M.R. (1999) *Sedimentology and Sedimentary Basins*. Blackwell Science, Oxford, 592 pp.

Leeder, M.R. and Perez-Arlucea, M. (2006) *Physical Processes in Earth and Environmental Sciences*. Blackwell Publishing, Oxford, 330 pp.

Miall, A.D. (2000) *Principles of Sedimentary Basin Analysis*. Springer-Verlag, New York, 616 pp.

Nichols, G. (2009) *Sedimentology and Stratigraphy*. Blackwell Science, Oxford, 355 pp.

Perry, C. and Taylor, K. (2007) *Environmental Sedimentology*. Blackwell Publishing, Oxford, 441 pp.

Potter, P.E., Maynard, J.B. and Pryor, W.A. (2005) *Mud and Mudstones*. Springer-Verlag, New York.

Reading, H.G. (ed.) (1996) *Sedimentary Environments: Processes, Facies and Stratigraphy*. Blackwell Science, Oxford, 688 pp.

Tucker, M.E. (2001) *Sedimentary Petrology: an Introduction to the Origin of Sedimentary Rocks*. Blackwell Science, Oxford, 262 pp.

Walker, R.G. and James N.P. (eds) (1992) *Facies Models – Response to Sea-Level Changes*. Geoscience Canada, 407 pp.

Warren, J.K. (1999) *Evaporites: Their Evolution and Economics*. Blackwell Science, Oxford, 438 pp.

Textbooks Dealing Specifically with Sedimentary Structures, including Trace Fossils

Allen, J.R.L. (1982) *Sedimentary Structures*, vols 1 & 2. Elsevier, Amsterdam.

Branney, M.J. and Kokelaar, B.P. (2002) Pyroclastic density currents and the sedimentation of ignimbrites. *Geological Society, London, Memoir*, **27**, 152 pp.

Bromley, R.G. (1996) *Trace Fossils: Biology, Taphonomy and Applications*, 2nd edn. Chapman-Hall, London, 361 pp.

Collinson, J.D., Mountney, N. and Thompson, D.B. (2006) *Sedimentary Structures*, 3rd edn. Terra Publishing, London, 207 pp.

Demicco, R.V. and Hardie, L.A. (1994) *Sedimentary Structures and Early Diagenetic Features of Shallow Marine Carbonate Deposits*, SEPM Atlas No. 1. Society of Economic Paleontologists and Mineralogists, Tulsa, 265 pp.

Hasiotis, S.T. and Van Wagoner, J.C. (2002) Continental Trace Fossils. SEPM Short Course Notes 52.

Jerram, D. and Petford, N. (2011) *Igneous Rocks in the Field*. John Wiley & Sons, Ltd, Chichester, 229 pp.

Leyrit, H. and Montenat, C. (eds) (2000) *Volcaniclastic Rocks: from Magmas to Sediments*. Gordon & Breach Science Publishers, Amsterdam.

McClay, K. (1991) *The Mapping of Geological Structures*. John Wiley & Sons Ltd, Chichester, 161 pp.

McPhie, J., Doyle, M. and Allen, R. (1993) *Volcanic Textures. Centre for Ore Deposit and Exploration Studies*, University of Tasmania, 197 pp.

Miller, W. (ed.) (2007) *Trace Fossils: Concepts, Problems, Prospects*. Elsevier, Amsterdam, 611 pp.

Noffke, N. (2008) The criteria for the biogenicity of microbially induced sedimentary structures (MISS) in Archean and younger, sandy deposits. *Earth-Science Review*, **96**, 173–180.

Pemberton, S.G., Spila, M., Pulham, A.J. *et al.* (2001) *Ichnology and Sedimentology of Shallow to Marginal Marine Systems*, Short Course Notes, vol. 15. Geological Association of Canada, 343 pp.

Potter, P.E. and Pettijohn, F.J. (1977) *Palaeocurrents and Basin Analysis*. Springer-Verlag, Berlin, 425 pp.

Retallack, G.J. (1997) *A Colour Guide to Palaeosoils*. John Wiley & Sons Ltd, Chichester, 175 pp.

Stow, D.A.V. (2005) *Sedimentary Rocks in the Field*. Manson Publishing, London, 320 pp.

Stratigraphic Procedure and Sequence Stratigraphy

Boggs, S. (2006) *Principles of Sedimentology and Stratigraphy*. Prentice-Hall, New Jersey, 774 pp.

Catuneanu, O. (2006) *Principles of Sequence Stratigraphy*. Elsevier, Amsterdam.

Catuneanu, O., Abreu, V., Bhattacharya, J.P. *et al.* (2009) Towards the standardization of sequence stratigraphy. *Earth-Science Reviews*, **92**, 1–33.

Catuneanu, O., Galloway, W.E., Kendall, C.G.St.C. *et al.* (2011) Sequence stratigraphy: methodology and nomenclature. Report for the ISSC. *Newsletter in Stratigraphy*.

Coe, A., Bosence, D., Church, K. *et al.* (2002) *The Sedimentary Record of Sea-level Change*. Cambridge University Press and Open University, 285 pp.

Hedberg, H.D. (ed.) (1976) *International Stratigraphic Guide*. Wiley Interscience, 200 pp.

Miall, A.D. (2000) *The Geology of Stratigraphic Sequences*, 2nd edn. Springer-Verlag, Berlin, 522 pp.

Rawson, P.F., Allen, P.M., Brenchley, P.J. *et al.* (2002) *Stratigraphical Procedure*. *Geological Society Professional Handbooks*, 57 pp.

Statistical Analysis of Stratigraphic Data (Cycles and Beds)

Burgess, P. (2006) The signal and the noise: forward modelling of allocyclic and autocyclic processes influencing peritidal carbonate stacking patterns. *Journal of Sedimentary Research*, **76**, 962–977.

Drummond, C.N. and Wilkinson, B.H. (1996) Stratal thickness frequencies and the prevalence of orderedness in stratigraphic sequences. *Journal of Geology*, **104**, 1–18.

Lehrmann, D.J. and Goldhammer, R.K. (1999) Secular variation in parasequence and facies stacking patterns of platform carbonates: a guide to application of stacking pattern analysis in strata of diverse ages and settings. In: *Advances in Carbonate Sequence Stratigraphy: Applications to Reservoirs, Outcrops, and Models*, Special Publication 63 (eds P.M. Harris, A.H. Saller and J.A. Simo). SEPM, 187–225 pp.

Murray, C.J., Lowe, D.R., Graham, S.A. *et al.* (1996) Statistical analysis of bed thickness patterns in a turbidite section from the Great Valley sequence, Cache Creek, northern Califormia. *Journal of Sedimentary Research*, **66**, 900–908.

Saddler, P.M., Osleger, D.A. and Montanez, I.P. (1993) On the labeling, length and objective basis of Fischer plots. *Journal of Sedimentary Petrology*, **63**, 360–368.

Weedon, G. (2003) *Time-Series Analysis and Cyclostratigraphy: Examining Stratigraphic Records of Environmental Cycles*, Cambridge University Press, Cambridge, 259 pp.

Useful Websites: here are a few to start with...

The British Geological Survey has much interesting and relevant information for students involved in fieldwork and mapping: http://www.bgs.ac.uk/education/resources.html

Ian West from Southampton University has created a site with many field photos, especially from southern England: http://www.soton.ac.uk/~imw/

The Society for Sedimentary Geology (SEPM) has a site with much information on facies, environments and sequence stratigraphy: http://sepmstrata.org/terminology.html

Try the Sedimentary Rocks section of the Geology Shop for a way into numerous websites: http://www.geologyshop.co.uk/sedime~1.htm

The US Geological Survey (USGS) site on bedforms is interesting: http://walrus.wr.usgs.gov/seds/

Chapter References

Chapter 1

Bouma, A.H. (1969) *Methods for the Study of Sedimentary Structures*. Wiley-Interscience, New York, 458 pp.

Bover-Arnal, T., Salas, R., Moreno-Bedmar, J.A. and Bitzer, K. (2009). Sequence stratigraphy and architecture of a late Early-Middle Aptian carbonate platform succession from the western Maestrat Basin (Iberian Chain, Spain). *Sedimentary Geology*, **219**, 280–301.

Chapter 2

Blackbourne, G.A. (2000) *Cores and Core Logging for Geologists*. Whittles Publishers, Caithness, 113 pp.

Catuneanu, O., Abreu, V., Bhattacharya, J.P. *et al.* (2009) Towards the standardization of sequence stratigraphy. *Earth-Science Reviews*, **92**, 1–33.

Tucker, M.E. (ed.) (1988) *Techniques in Sedimentology*. Blackwells, Oxford, p. 408. See Chapter 2 by John Graham, 408 pp, on field techniques.

Zervas, D., Nichols, G.J., Hall, R. *et al.* (2009) SedLog: a shareware program for drawing graphic logs and log data manipulation. *Computers & Geosciences*, **35**, 2151–2159.

Chapter 3

Jerram, D. and Petford, N. (2011) *Igneous Rocks in the Field*. John Wiley & Sons Ltd, Chichester, 229 pp.

McPhie, J., Doyle, M. and Allen, R. (1993) *Volcanic Textures. Centre for Ore Deposit and Exploration Studies,* University of Tasmania, 197 pp.

Chapter 4

BSI (British Standards Institute) (1981) *Code of Practice for Site Investigations*. BS 5930, 140 pp.

Chapter 7

Potter, P.E. and Pettijohn, F.J. (1977) *Palaeocurrents and Basin Analysis*. Springer-Verlag, Berlin, 425 pp.

Chapter 8

Catuneanu, O., Galloway, W.E., Kendall, C.G.St.C. *et al.* (2011) Sequence stratigraphy: methodology and nomenclature. Report for the ISSC. *Newsletter in Stratigraphy*.

Sea-stack (20 metres high) and cliff of Permian slope-apron carbonates with large round concretions. Marsden, NE England.

INDEX

Opposite: A Wulff net for the reorientation of palaeocurrent data collected from rocks dipping at more than 30°. This figure is also available on the internet at www.wiley.com/go/sedimentaryrocks4e.

Also available on the internet are the following figures from this book:

Logging sheet for the field (front inside cover)
Logging sheet for cores
Symbols for graphic logging (Figure 2.2)
Recording sheet for palaeocurrent readings and rose diagram (Figure 7.2)

www.wiley.com/go/sedimentaryrocks4e